T0179712

Why Democracies Need Science

Why Democracies Need Science

Harry Collins and Robert Evans

polity

The right of Harry Collins and Robert Evans to be identified as Authors of this Work has been asserted in accordance with the UK Copyright, Designs and Patents Act 1988.

First published in 2017 by Polity Press

Polity Press
65 Bridge Street
Cambridge CB2 1UR, UK

Polity Press
350 Main Street
Malden, MA 02148, USA

ISBN-13: 978-1-5095-0960-7
ISBN-13: 978-1-5095-0961-4 (pb)

A catalogue record for this book is available from the British Library.

Library of Congress Cataloging-in-Publication Data
Names: Collins, H. M. (Harry M.), 1943- author. | Evans, Robert.
Title: Why democracies need science / Harry Collins, Robert Evans.
Description: Cambridge, UK ; Malden, MA : Polity Press, [2017] | Includes
   bibliographical references.
Identifiers: LCCN 2016038437 (print) | LCCN 2016055612 (ebook) | ISBN
   9781509509607 | ISBN 9781509509614 (pb) | ISBN 9781509509645 (Epub)
Subjects: LCSH: Democracy and science. | Science--Political aspects. |
   Science and state.
Classification: LCC JC423 .C6478 2017 (print) | LCC JC423 (ebook) | DDC
   338.9/26--dc23
LC record available at https://lccn.loc.gov/2016038437

Typeset in 11 on 13 pt Bembo by
Servis Filmsetting Ltd, Stockport, Cheshire
Printed and bound in Great Britain by CPI Group (UK) Ltd, Croydon

For further information on Polity, visit our website:
politybooks.com

Contents

Preface

There are four parts to our argument. Part I introduces the problem, setting out the main issues as we see them, and describes the academic foundations on which our call to arms is built. Part II contains most of the new ideas: it sets out the principles that inform what we call 'elective modernism' and explains their implications for the ways in which scientific advice should be sought and used in policy-making. We argue that science should be seen as a moral enterprise and that the values that inform scientific work should be celebrated; this, as far as we know, is a new idea, so it takes precedence over the utilitarian justification of science in the argument of the book, but it can also be used in addition to the utilitarian argument when that works. Crucially, however, the moral argument works for sciences that do not have any obvious utility and, in that sense, it is prior. We argue at the same time for the primacy of democratic institutions in technological decision-making and invent a new kind of institution – 'The Owls' – whose job is to represent faithfully the content and degree of certainty of any technical advice that might be thought to bear on these decisions. Part III shows that we do what we say should be done in Part II. There we suggest that one of the values that characterizes science is 'continuity', by which we mean that even the most revolutionary of scientific ideas will seek to incorporate and retain a good portion of what was previously accepted as true. In Part III, we show the ways in which our ideas, which we have come to realize in the light of reactions to them must include an unintended element of revolutionary thinking, relate to the huge existing literature that deals with science and democracy. In Part IV, we sum up our argument in a manifesto for the future of science that sets out the key choices facing you, the reader, in as straightforward and uncompromising a manner as possible. Given what has been said so far, it will be no surprise that this manifesto emphasizes the moral responsibility of scientists to act in ways that preserve science's traditions and values. If scientists fail in this task and we fail to support them in it, then a crucial element of the culture that sustains democratic societies will be lost.

Though both authors take full responsibility for the whole of this book, Collins was the lead author of Part II while Evans took the lead on Part III. The authors have to thank many people. We thank Martin Weinel for his marvellous analysis of the Thabo Mbeki, anti-retroviral drugs affair and for his contributions to the more political parts of this book. Under slightly changed circumstances, he would have been a co-author. Above all, we thank the various audiences who have been willing to listen to talk of elective modernism. The term had been batted around a bit but the ideas were probably first presented by Collins on 8 October 2008 at the regular meeting of Cardiff's Centre for the Study of Knowledge Expertise and Science, and since then they have been presented at many national and international meetings and mentioned, *en passant*, in a few pieces of published work. Intervening events have slowed their presentation in extended form much more than we anticipated.

Part I

Introduction

1

Science as a Moral Choice

What kind of society do we want to live in? There is plenty wrong with Western societies: huge and growing inequalities; unstable and corrupt financial systems; political systems whose logic places national self-regard above the terrible suffering of distant nations; and politicians for purchase. Worse, in Western societies we are no longer confident about our basic values. The realization that a sense of moral superiority was often a thin disguise for the exploitation of colonized peoples, and now the fear that exploitation of the Earth's natural resources is risking our collective future, are causing us to question what we have traditionally thought of as progress. Maybe the exploited peoples had it right and a calm life in tune with nature, however short, and however bereft of technological goods, is better than the endless quest for more, for further, for faster.

This book deals with questions one level down from these concerns, taking it that the difference in quality between our lives and those of our distant ancestors does represent progress. There has been material progress, such as freedom from high rates of maternal and infant mortality and relief from the struggle to eat and stay warm, and, to a less marked extent, there has been moral progress, with the weak no longer living in continual fear of the strong. The problem that we address here is the potential or actual erosion of our style of life which is coextensive with the erosion of certain once-cherished values. We address one part of this problem: the role of science in society.

In earlier works – discussed later under the heading 'Three Waves of Science Studies' – we argued that, in spite of the huge enrichment of our critical understanding of the nature of science that has taken place since the 1970s, it is still important and intellectually possible to value expertise. Big arguments can be based on shallow foundations and we base our ideas about expertise on the commonsense view that it is better to give more weight to the opinions of those who, literally, *know* what they are talking about. But there are all kinds of experts who know what they are talking about: astrologers and astronomers; chemists and alchemists; tea-leaf readers and econometricians. In our earlier arguments, there is only a brief justification of scientific expertise as opposed to other kinds of expertise. Here we complete another step in the Third Wave project by justifying scientific expertise in particular while making it as hard for ourselves as possible by accepting pretty well everything from the social constructionist critique of science that has emerged since the 1960s – The Second Wave. These days, academic discussions of science and technology grow ever more polarized, but the view here fits neither pole; it endorses the enriched understanding and critique of old models of science that came with the cognitive revolution of the 1970s but it also aims to preserve a special place for science in society.

The moral case for scientific values

We are interested in problems that can be understood in terms of the shared values and practices of social groups. As particular practices are repeated over time and become more widely shared, the values that they embody are reinforced and reproduced and we speak of them as becoming 'institutionalized'. In some cases, this institutionalization has a formal face to it, with rules and protocols written down, and specialized roles created to ensure that procedures are followed correctly. The main institutions of state – parliament, courts, police and so on – along with certain of the professions, exhibit this formal character. Other social institutions, perhaps the majority, are not like this; science is an example. Although scientists are trained in the substantive content of their discipline, they are not formally instructed in 'how to be a good scientist'. Instead, much like the young child learning how to play 'nicely', the apprentice scientist gains his or her understanding of the moral values inherent in the role by absorption from their colleagues – socialization.[1] We think that these values, along with the values that inform many of the professions, are under threat, just as the value of the professions themselves is under threat.

The attacks on science come from many sources. From the outside, science is beset by post-modernist analysis that sees no truth, only 'accounts'; it is beset by environmentalist critiques that see science as an instrument of ecological disaster; and it is beset by political regimes that see value only in economic terms, or, in America, can make political capital by contrasting science unfavourably with religion. Even in our own subject – the social studies of science – one never hears an argument or a position defended on the grounds that it is 'scientific'; the very idea would be dismissed as naïve since it is now believed there no longer is such a thing as science distinct from society. Science is also under attack from the inside. Scientists, thinking to defend their culture from politicians wishing to reduce taxes, rush to embrace the idea that they can deliver material and cultural goods to society – science is in there with capitalism forging new start-up companies, providing impactful outputs that increase productivity and efficiency, and entertaining the masses with astonishing revelations about the nature of the heavens. But you need a long spoon to sup with the devil. The danger is that soon science will be valued only for its material and entertainment value. The intention may be good but too many scientists are selling their profession in the wrong marketplace.

Professions, professionalism and moral leadership

A society is made up of institutions: transport systems, educational systems, healthcare services, providers of housing, food producers, police, lawyers, the military, sportspersons, entertainers, churches, political institutions, businesses and banks. The moral life of a society is, in part, an aggregate of the moral substance of these institutions. In institutions like religion, the moral role is explicit. But religion is also the most obvious example of how the moral leadership role of an institution can decline. In the UK, the established church – the Church of England – is still saying all the right things, but hardly anyone is listening. In the US the situation is different, with religious institutions still strong, but there are many competing ideas, very few of which are ready to confront the dominance of free-market capitalism. And it is probably free-market capitalism that has had the most corrosive influence on democratic life in the second half of the twentieth century, not least because it has subverted and undermined the notion of professionalism.

In some of the earliest work on the nature of professions (e.g., by Durkheim and later Parsons[2]), professions such as law and medicine are explicitly linked with the moral qualities expected of their practitioners and the stabilizing effect this had on society as a whole. In contrast, the contemporary idea of professionalism has a more managerial and ideological meaning in which ideas of autonomy and personal responsibility are used to retain some degree of market power but also, within organizations, to discipline workers by creating normative expectations of duty, responsibility and care. This marketization of the professions, in which professionalism 'becomes more commercially aware, budget focused, managerial, entrepreneurial and so forth' undermines the idea of professions as repositories of moral standards.[3]

In many spheres of work, these changes in work practices are clearly visible. Professionalism is widely trumpeted as a value for workers of all sorts, and new professional bodies spring up all the time to protect this new jurisdiction. Whilst, for many, the contemporary demands of professionalism are experienced as attempts by those in more senior positions to devolve responsibility to those lower down the organizational hierarchy, for those with genuine autonomy there is evidence that the old moral codes no longer apply. When one of the authors of this book was young, the banks could be held up as an object lesson in integrity. The success of 'The City' – London's 'square mile' – was said to be based on the fact that everyone knew that a handshake could seal a deal that would never be broken. Collins's mother told him about her friend, who worked in a bank, once spending the entire night searching for the mistake that had caused a discrepancy in the accounts amounting to a halfpenny. But Thatcher's doctrines – 'greed is good', and 'there is no such thing as society' – backed up by Reagan's free-market religion, led to Enron, to crash after crash, and a succession of corruption scandals such that, nowadays, in so far as the banks offer leadership, it is in unbridled self-interest.

In the UK it sometimes feels as if there are now hardly any institutions that the citizen can trust: politicians fiddle their expenses, celebrities turn out to be sex offenders, newspapers hack into the voicemails of private citizens to source stories, energy companies have tariff structures so complicated it is impossible for consumers to make good choices, sports administration is corrupt and performances aided by organized regimes of doping, food labelling is no longer accurate and so on, with a new revelation every week or so. Ironically, the time now required to check and re-check is, in economic terms, colossally inefficient, just as living in some regions of the developing world can be colossally inefficient because of the day-to-day corruption against which market theorists rail. The value of self-interest that the religion of the market promotes is nowadays driving our societies backwards just as once it drove them forward.

Notice, then, that the concern of this book is out of kilter with much contemporary social science: we are not attempting to solve problems of inequality or inter-generational justice. We agree these problems are serious but so much effort is already directed at them that we risk producing a sociological monoculture. In contrast, we are concerned with preserving the fragile tissue of democratic norms and values that is being eroded by the day-to-day violence, corruption and crude exercise of government-sponsored force in many nations around the world and by the growing corrosion of our own 'Western' societies, driven by an unconstrained free-market ideology. In contemporary science and technology studies, the predominant motif is to eliminate the division of powers between science and politics in order that science and technology can become socially responsible. In contrast, our motif is to safe-guard the division of powers so that science and technology can act independently of society! Most social analysts think that democracy needs protecting against scientific and technological experts; we argue that scientific and technical experts have the potential to protect democracy!

The difference arises out of what you think about society: if you think our existing societies are benign, then it may be wise to make science and technology answer to them, but if you think our societies are becoming more corrupt and less benign, then you might want science and technology to retain their independence. The principle will be familiar to the academic readers of this book who insist that their own independence of thought be protected by university tenure or its equivalent. Indeed, it is quite striking that university academics are so strident in their justifications of tenure, while at the same time many demand that science answers to the demands of society. We base our argument around the norms of science and it is not surprising that the best-known previous discussion of these norms was associated with the rise of fascism – with the sudden realization that societies were not going to be benign for much longer, so independence for scientists and academics in general became what we desperately wanted. That is why we are looking to science as an institution that can give moral leadership, rather than as something from which society needs protecting.

Science is not the only institution that has the ability to hold the moral line. In the UK, the National Health Service is another – at least, bits of the National Health Service (NHS). Both authors can attest that when you make it to the head of a waiting list, or if you are unlucky enough to need emergency treatment, the NHS is brilliant at every level, from the consultants to those who empty the bed pans. Consultants, of course, are well paid, but the middle ranks of nurses and auxiliaries are not, yet, in our experience, they provide a level of care, both emotional and practical, that leaves one knowing there still is such a thing as society and that contact with lives lived with integrity is rewarding beyond riches. The trouble is, of course, that mostly the NHS is celebrated for long waiting lists and newsworthy scandals; the NHS is not in a position to give moral leadership to anyone but the seriously ill and they are not a particularly vocal minority – especially as the rich and powerful are more and more sucked out of its domain and into private care.

Are there other flourishing institutions providing moral leadership for Western societies? It is hard to think of any obvious cases. What is sure is that we need such institutions desperately. What we argue in this book is that science could be one of these institutions – an institution that can provide moral leadership. This is because *good* actions are intrinsic to science's *raison d'être*. It has become the fashion to attack science and, of course, we have seen the corrosive effect of marketization on science, not only in the way honest scientists are persuaded to sell their profession but also with increasing fraud and the willingness of some 'scientists' to adjust the content of their findings according to the price on offer. Nevertheless, parts of science are still intact. We need to make science's special nature clear, and show society what it stands for, before it is overwhelmed by the free-market tsunami like so much else.[4] This is not a worldly-wise book, it is not smartly sophisticated; it is desperately grasping for the last vestiges of naïveté.

Three Waves of science studies

The academic foundations of this book are the social studies of science, but it takes a stance that is at variance with its mainstream. It is useful to divide the social studies of science into three waves. As with all heuristics, the categorization is not perfect and, as we demonstrate, there is some continuity and overlap between the waves. Despite this the idea of three waves provides a quick and easy way of setting out the key issues and why they matter.

To start at the beginning, Wave One was the period in which it was believed that science was unquestionably the pre-eminent form of knowledge-making and that its knowledge was absolute and universalistic. Wave One runs from at least the early twentieth Century, if not earlier, and was at its most influential in the 1950s and early 1960s, when Lewis Strauss, then chairman of the United States Atomic Energy Commission, predicted a future in which 'our children will enjoy in their homes electrical energy too cheap to meter'.[5] Within the social sciences, this was the period in which Robert Merton was writing about the importance of democratic societies fostering scientific norms (see chapter 2), and in which the social analysis of science focused on explaining scientific error not scientific truth. Under Wave One, the correctness of scientific research needed no social explanation – it was true, so no further explanation was required – but what did need to be explained, typically by social mechanisms such as prejudice, bias, special interests and so on, was how false beliefs were mistakenly taken to be correct. Although this view is no longer supported by many in the social sciences, it remains the commonsense view of practising scientists. It also informs many popular representations of science including, for example, the forensic science genre of police procedurals in which technical analysis inexorably reveals the facts of the matter.

Wave Two is more recent, with what are now seen as its foundational works – in particular, Thomas Kuhn's *Structure of Scientific Revolutions* – emerging in the 1960s. It would be wrong to say that all these early authors were engaged in, or even supported, the social constructionist analysis of scientific knowledge in which both 'true' and 'false' beliefs were explained in the same way. Instead, this period saw the publication of several major works, of which Kuhn's is the exemplar, that were taken up by others and used as the basis of analyses that showed that scientific truth is best seen as an outcome of negotiation and agreement located within social groups.[6] Over the 1970s, 1980s and 1990s, Wave Two social scientists produced a wide range of case studies that demonstrated that the scientific method could not work as previously advertised and that scientific findings were much more affected by their social context than had previously been believed. This, in turn, had important implications for the role of science in society and, in particular, the use of scientific advice in policy-making. In brief, Wave Two provided a powerful argument against technocracy by showing how expert advice rested on a sea of social assumptions. This, in turn, led to arguments in favour of the democratization of science, and of expertise more generally, in order that the interests and priorities that inevitably coloured expert advice would better reflect the concerns of the wider society. Collins is a founding contributor of Wave Two and continues to work in this vein to this day (e.g., in his studies of gravitational wave physics).[7]

Wave Three, of which this book is a part, accepts everything that Wave Two has said about the nature of scientific work but disagrees with its conclusions. Where Wave Three differs from Wave Two is in its normative position. Wave Two has a 'default setting' in favour of more democratization, but Wave Three wants to replace this with a dial that can be turned to different places according to the nature and integrity of the science. The aim of Wave Three of science studies is to preserve the idea of expertise as specialist knowledge and to find a better way of analysing and managing the trade-offs between expert authority and democratic accountability. In what follows, we briefly outline the contrasts between the Wave Two and Wave Three approaches. We then outline the concepts and ideas that have emerged from this work and explain how they inform this book.

Elements of the Third Wave

Wave Three begins with a paper that we published in the leading science studies journal *Social Studies of Science* (Collins and Evans, 2002). That paper contained both a technical element, expressed as a nascent classification of expertise, and a political element, expressed as a call for scholars in Science and Technology Studies (STS) to use their expertise about expertise to intervene in public debates. In particular, we argued that STS should use its unrivalled understanding of knowledge-making practices to help inform decisions about whether or not particular groups or individuals have a legitimate claim to expertise.

The paper has now been cited many times (1,700+ by June 2016) and, in the months following its publication, it received four published replies – one positive but three negative – to which we were allowed to publish a formal response.[8] Although the critical replies made their arguments in different ways, they shared a general sense that the political argument of the 'Third Wave' was a step backward rather than forward. It was suggested that it was incompatible with Wave Two rather than consistent with it, and that it presaged a return to technocracy by giving undue power to scientific experts. Despite our many carefully worked-out denials of this claim – for example, we state over and over again that democracy always has the last word and that the only thing we ask is that it not misrepresent the claims of experts – this misrepresentation of *our* position regarding the political implications of the Third Wave approach is ever present (see chapter 4). In contrast, the technical element did not receive anything like the same degree of criticism and has been much less controversial.[9]

Following the initial publication, much of our work concentrated on the technical part of the Third Wave programme, leading to a much richer classification of expertise than originally proposed, a lot of work on the idea of interactional expertise, and the development of the Imitation Game as a research method.[10] As a result of this effort, the political side of the programme remained relatively neglected until the (as we saw it) inaccurate portrayal of the Third Wave by Frank Fischer in his book *Democracy and Expertise* (2009). This led to an exchange of papers in the journal *Critical Policy Studies*, of which 'The Politics and Policy of the Third Wave' represents our attempt to lay to rest some of the misunderstandings created by the initial responses to the Third Wave paper. Unfortunately, the replies to this paper suggest that we were not entirely successful. Our response to these replies was thus yet another attempt to set the record straight, with this volume providing, in addition to its main aim, the complete – and we hope compelling – statement of our position needed to reassure critics that our views do not lead in the dangerous directions they say.[11]

Problems of legitimacy and extension

One way of understanding the difference between Wave Two and Wave Three is to see them as attempts to solve different problems created by technological decision-making in the public domain. Wave Two, at least in its political guise, is an attempt to solve the 'problem of legitimacy' that arises when expert authority is allowed to ride roughshod over the concerns of others. Wave Two provides both a diagnosis and solution for this problem by demonstrating: (a) that the apparently neutral and objective advice provided by technical experts cannot have the unquestionable epistemological authority it claims; and (b) that a more robust solution could be reached by incorporating a wider range of perspectives and experiences into the decision-making process. A simple example, taken from Alan Irwin's *Citizen Science*, makes the point: in the late 1970s and early 1980s, farmworkers in the UK were concerned about the safety of an organophosphate herbicide called 2,4,5,T. The farmworkers believed that it caused a number of health problems, including miscarriages and birth defects, whereas the government advisors, formally represented by the Advisory Committee on Pesticide (ACP), insisted that the chemical was safe as long as it was used correctly. In reaching their decision, the ACP made an epistemic judgement that the experiential evidence of individual farmworkers was worth less than epidemiological studies or laboratory research, and a social judgement that the facilities and training needed to use the chemical safely were routinely available to farmworkers. This last point is particularly important as one might reasonably expect that the farmworkers would be better informed about the circumstances in which 2,4,5,T was actually used than the scientists on the ACP. The implication of Irwin's analysis – and it is surely correct – is that the ACP's decision would have been more robust – and certainly more legitimate – if it had deferred to the farmworkers in those areas where they had the relevant experience and recognized that the normal conditions of use fell some way short of the recommended standards.[12]

In contrast, Wave Three is concerned with a different, albeit related, problem: the problem of extension.[13] The problem of extension arises because the arguments put forward under Wave Two did not contain any explicit criteria for setting limits to participation: Wave Two opens the policy-making door to new kinds of expert but has no mechanism for determining who should be granted entry and who should remain outside. But consider cases such as 2,4,5,T, where the extra knowledge needed can be found only among certain specialists who happen not to be qualified scientists but who still have privileged access to a particular kind of experience-based expertise. Huge confusion was caused by advocates of more democratization referring to experts such as the farmworkers as 'lay experts', because it made it seem as though anyone could be an expert. The problem of extension is thus concerned with how to operationalize 'more heterogeneous participation' in such a way that the relevant expertises identified under Wave Two approaches are included, but irrelevant non-expert contributions are excluded from the expert forums that feed into democracy. If done correctly, the solution to the problem of legitimacy is also the solution to the problem of extension: all the 'right' people will have a say in the technical debate, and those who have no relevant specialist expertise will contribute as citizens participating in existing democratic institutions without pretending to be, or being described as, experts.[14]

A typology of expertise

In order to define what 'relevant specialist expertise' means, Wave Three begins by treating expertise as 'real' – that is, as something a person or group possesses – and then develops a theory of expertise that can be used to distinguish between different types, levels and kinds of expertise.[15] This is the technical element of the programme. The classification of expertises is explained in most detail in *Rethinking Expertise*, from which table 1.1 is taken. It is based on the sociological axiom that expertise is the outcome of successful socialization into a social group. The various skills and expertises an individual possesses are then the accumulation of the social groups in which he or she is a successful participant, while the absence of socialization implies the absence of that expertise.[16]

The structure of the table is founded on the different ways in which it is possible to participate in a social group. Working from the top, the first two rows identify the society-wide *ubiquitous expertises* and *dispositions* (personal characteristics) that enable socialization to take place, and which lay the foundation for developing narrower *specialist expertises* and more generic *meta-expertises*.

Table 1.1 The Periodic Table of Expertises

UBIQUITOUS EXPERTISES			
DISPOSITIONS	Interactive ability		Reflective ability
SPECIALIST EXPERTISES	**UBIQUITOUS TACIT KNOWLEDGE**		**SPECIALIST TACIT KNOWLEDGE**
	Beer-mat knowledge	Popular understanding	Primary source knowledge	Interactional expertise	Contributory expertise
				Polymorphic	Mimeomorphic
META-EXPERTISES	**EXTERNAL**		**INTERNAL**
	Ubiquitous discrimination	Local discrimination	Technical connoisseurship	Downward discrimination	Referred expertise
META-CRITERIA	Credentials		Experience		Track-record

(*Source*: Collins and Evans, 2007)

The different types of specialist expertise correspond to commonsense understandings. Thus, the first three categories denote the kinds of understanding that can be achieved solely by using resources such as magazines, books, academic journals, Google, YouTube videos and so on. As these do not permit any direct interaction with the community in question, none of the tacit knowledge that is unique to that expertise can be acquired by the learner; instead, the learner has only the ubiquitous tacit knowledge needed for everyday life, bolstered, perhaps, by specialist *information*. In contrast, the last two types of specialist expertise – interactional expertise and contributory expertise – do require immersion in the relevant community and so enable the learner to develop the specialist tacit knowledge used in that domain of practice. Thus, in the example given above, we might say that farmworkers had contributory expertise in the practical application of organophosphate herbicides in outdoor environments and, on that basis, had a legitimate contribution to make in discussions about the regulation and use of these chemicals.

The meta-expertises row captures the opportunity cost of the extended socialization needed to become a contributory expert and identifies the ways in which people can make judgements about expert claims even though they are not experts themselves. Again, there are some methods – ubiquitous and local discrimination – that do not depend on any knowledge of the domain in question and which 'transmute' purely social judgements about who to trust into beliefs about the state or nature of the world. There are other abilities – technical connoisseurship, downward discrimination and referred expertise – that differ in that they require some familiarity with what is being judged.

Technical and political phases

The second element of the Third Wave paper was the more political one. Here we argued that STS could contribute to technological decision-making in the public domain and that doing this required distinguishing between two kinds of activity. These were labelled the 'technical phase' and the 'political phase'. The intention was to draw attention to the ways in which technological decision-making combined two different institutional practices, and to argue that it was important to retain the distinction between them if science, already seen by some as 'politics by other means', was not to become indistinguishable from the networking, horse-trading and pork-barrelling that characterize the overtly political elements of the public domain. Much of the controversy about the Third Wave paper turns on this difference, with critics claiming that it is impossible to distinguish between the technical and political without reinventing the Wave One fact–value distinction that Wave Two had so comprehensively destroyed. As this entire book can be seen as a rejection and refutation of this claim, we do not dwell on it here and, instead, concentrate on summarizing the distinction as it was originally proposed and noting the relatively minor changes that have emerged since then.

Distinguishing between the technical and the political phases breaks the idea that policy decisions should be informed by the best expert advice into its constituent parts – 'decision' and 'advice' – and allows us to argue that each part needs to be approached in a different way and held accountable to different standards. The technical phase refers to the soliciting of expert advice and is characterized by two features. The first, which links it to Wave Two, is that expert advice should include experience-based experts from outside science wherever this is relevant. It recognizes the problem of legitimacy by saying that there are many kinds of expertise, but also recognizes the problem of extension by saying that domains of expertise should always be represented by interactional or contributory experts. The second feature of the 'technical phase' is that it is characterized by 'scientific' norms in that its politics should be 'intrinsic' rather than 'extrinsic' – in other words, the normative claim is that technical advice should be provided in such a way as to minimize the effect of political bias and influence. This means that technical advisors should not exaggerate the certainty or impact of their work in order to influence the political decision, and that decision-makers should not demand that engineers and scientists draw firm conclusions when this cannot be done. Of course, we know that Wave Two shows that a complete separation of facts and values is impossible in practice: the rationale for insisting on it anyway is that, like perfect justice, perfect democracy and other unattainable aspirations, the value of the goal is not diminished by the difficulty of achieving it.

The political phase referred to the social institutions and processes that deal with this expert advice and arrive at a policy decision after taking it into account. Here the normative claim was that the political phase would run on democratic principles and that all citizens, regardless of their expertise, would have an equal stake. Crucially, the political phase always has priority over the technical phase and so there is no question of technocracy. It is also entirely possible, and consistent with what has been said so far, that democratic institutions will differ in their forms and procedures and may also differ in the decisions they take even if given the same expert advice.

Since the 2002 paper, there have been some minor changes to the detailed arguments associated with how these principles can be applied, but the underlying distinction remains essentially unchanged. Some minor changes that occurred before and during the writing of this book can be summarized as follows:

1. A more formal analysis of the relationship between the expertise needed for meaningful participation and forms of democratic participation has led to a general preference for more deliberative forms of democracy.[17]
2. A clearer statement of the relationship between the technical and political phases – dubbed 'the minimal default position' – has been developed. It says that political decision-makers should not misrepresent technical advice in order to present political choices as technical necessities.[18]
3. A more elaborate analysis of the different ways in which contributions to the technical phase could be made has been set out.[19]
4. We have recognized that it is important to distinguish between democracy and politics: democracy refers to a particular set of political principles and institutions that we largely support and which are entirely compatible with science; politics, in contrast, is the more visceral work of building alliances and accumulating the power needed to get things done.

Revelation versus proof

Wave Two was a revolution in our understanding that placed science in a very much more intimate relationship with politics and non-scientific interests in general: it made it seem that there was nothing special about science. We take it that what was found out under Wave Two of science studies makes it very difficult to defend science on the grounds of its truth and utility. Instead, we argue that democracies need science because science is, or can be, a fountainhead of good values. Science's understandings are continually disputed, but science's values are eternal. Thus, whatever position one takes on the value of science's outputs – and most people will still reach for truth and utility when justifying science – the argument from science's values will stand – as long as it works. Democracies, assaulted by the forces of free-market capitalism, are in desperate need of a fountainhead of values. We argue that science has the potential to provide it. But how do you justify a choice of values?

Imagine: you are arguing with a casual acquaintance who says that he is intending to torture some children; he is going to do it for no particular reason – he is going to do it gratuitously. He explains that he has studied moral philosophy and can find no decisive proof that the gratuitous torture of children is wrong. In fact, he can find no decisive proof of the rightness or wrongness of any act at all and in future he intends to act on whim and independent of what are normally taken to be moral prescriptions. He won't torture the children, however, if you can provide a decisive proof that it is an evil thing to do.

Now, it is indeed the case that moral philosophy has not provided a decisive proof that acts that we generally consider abhorrent really are evil. Faced with this casual acquaintance, is closing the gap in the countless volumes of moral philosophy the right way to stop him torturing the children? No! First, it is likely to be a hopeless task; and, second, one already knows that something has gone wrong with a person who thinks that the way to make moral choices is with decisive proofs. It is a characteristic of moral action that we make such choices without seeking decisive justification. Philosophers aside, philosophical justifications are not the source of moral judgements. Given someone who is intending to act in a gratuitously amoral way because he or she thinks that decisive proof is the crucial thing, whatever the right course of action is, it is certainly not another philosophical argument.

In much the same way, what we will call 'elective modernism' takes science to be a matter of moral choice: the word 'elective' implies choice; the word 'modernism' has to do with science.[20] Science is what we *should* choose as our approach to understanding the physical, biological, psychological and social world – in short, the observable world. We just know when it comes to the observable world, those who have observed in a systematic way are a *better* source of opinions than those who have not. Note the word 'better', which is not the same as 'more correct'. It is just better in the sense of goodness, rather than correctness. We value people who have observed for their observations; to have observed is better than not to have observed. We say that such people have experience and expertise. The high value given to expertise is central to elective modernism just as it is central to science itself but, as we will see, there are many other values associated with science and with elective modernism. As with ethical choices, we know that we must choose expertise over no expertise, and we must choose expertise informed by scientific values even if we do not think we know how to justify the choice in any foundational way. To repeat, not being able to reach that kind of foundation means the italicized '*better*' in the earlier sentence cannot mean 'more efficacious' – if it did we would have a foundational justification; it does not mean better *at anything*, it just means better.

In countless volumes, philosophers of science attempt to prove that science is best, but they never quite succeed – just as the moral philosophers never quite succeed. What is done here is to make the nature of the choice and its alternatives as clear as possible in the hope that the right choice will be persuasive to the point of self-evidence. To a large extent the defence of science on moral grounds, like other moral arguments, consists of helping us notice how we already act and think. In that sense, what is going on here is more in the nature of a *revelation* than an argument. At other times, what we are doing is perhaps best described as offering a *proposition*. The attractiveness of the proposition is reinforced by setting out the alternatives – the alternatives cannot be proved to be abhorrent but if they do not seem immediately abhorrent to you then there is something wrong with you – in the same way as there was something wrong with the person who was going to torture children gratuitously. If you cannot see what is wrong, the book will find no purchase.

Procedures not facts

So far, two things have been said:

1. Science can be *revealed* to be 'good', in a moral sense, when it comes to the approach to the observable world.
2. On close examination, other kinds of justifications or defences of science, such as philosophical or utilitarian defences, often fail.

It is the first of these that is the most important. Though the authors believe the second statement, readers do not have to believe the second to accept the first. Please, dear reader, do not abandon this book yet just because you are convinced that science can be justified in a 'rational' way. You do not have to believe what is written next to persevere with the book.

Over the last fifty or more years, the work of philosophers, sociologists and historians of science has given rise to a much deeper and more detailed understanding of science. They have shown that science's day-to-day practice is not – and, it has been argued, cannot be – driven by a well-defined logic. They have also argued that science cannot be defended on the grounds of its success. To repeat, this is what we called the Second Wave of science studies. During the First Wave, philosophers and sociologists of science took it to be their job to explain how science – the self-evidently pre-eminent form of knowledge – worked, and how society could be arranged to nurture it. The Second Wave took science to be just as open to the analysis of the sociology of knowledge as any other form of knowledge; it showed that science was not so special after all and it levelled the epistemological terrain and, as it became embroiled in the wider post-modern movement, it tended to devalue science.

The Third Wave, of which, as we have said, this book is a manifestation, accepts this major change in our understanding of science but tries to show that, in spite of the many brilliant findings of the Second Wave, there are still ways to value expertise and to value science; the epistemological terrain, it is argued, has not been irrevocably levelled. The Third Wave sees the analysis of the Second Wave as correct but the normative conclusions drawn from it as incorrect. We argue that those of us, including the authors, who have worked in the Second Wave, or who accept its conclusions, are in great need of a moral defence of science. Post-modernism never had more than one shot in its locker – a sceptical shot. It was a powerful shot and brought down the strong and geometrically perfect ramparts of science revealing the higgledy-piggledy streets of the city behind. But one cannot continue to build upon scepticism; it brings things down but does not built them up again.[21] If we want to move forward, we need a post-post-modernism. Elective modernism is our candidate because it can move things on without abandoning the higgledy-piggledy but vital city and without rebuilding the old defensive walls. Elective modernism offers us the chance of living in the city of science without abandoning the new understandings of science developed over the last half-century.

To repeat, pleadingly, even if you are one of those who are unconvinced by the work of the Second Wave – someone who has always dwelled and continues to dwell in the city without accepting that the walls have come down or that the streets do not run straight – you should still be interested in the attempt to defend science as a moral choice. This is because, if the moral defence works, you will have two strings to your bow. You can use the moral string should you ever have to confront the possibility that the 'truth and utility string' is looser than you thought. Elective modernism rests its case on what it is to be scientific, and ignores outcomes and facts; it is *scientific values* that are going to be said to be central to our culture not *scientific facts and outcomes*. If good science also produces important outcomes, then that is a bonus and, to repeat, if that is what you think justifies science, then this book just adds a second string to the justificatory bow, while those who don't believe it can still shoot arrows justifying the centrality of science in our lives with the string we provide.

It is argued here, furthermore, that, even for the convinced rational defender of science, it is *safer* to rely on scientific values than scientific outcomes in defending science because resting one's case on scientific facts gives a hostage to fortune – scientific findings believed by some are not believed by others, and what is believed now may not be believed in the future. Another reason is that in the aftermath of the atrocities committed during the Second World War it was feared that science was responsible for replacing moral worth with calculation – a strong technocracy. Too much respect for science's efficacy can lead to such things if the gaze becomes too narrow. Respect for science's values has none of these dangers, more especially since the programme is based on the disrespectful, inexact, untidy model of science – the higgledy-piggledy city – to which the sociology of scientific knowledge (SSK) and related approaches gave rise. No one immersed in this model of science is going to use it to justify mass murder – not even mass murderers are going to use it this way.[22]

Structure of the book

This is a book about science as a moral institution. As explained, there are things about the book that will be unfamiliar to the community of academics in which we work. For example, right at the heart of the book, just where the academic would normally look for justification, there is nothing! The book is built on even less than our commonsense argument about expertise: it is built on the self-evident goodness of certain moral beliefs and actions. It is not that we have forgotten to justify our position, it is just that when everyone has finished arguing there is nothing left but this: you just know that the gratuitous torture of children is wrong, and so it is with observation and the other scientific values that we will set out.

The *choice* of science as our way of understanding the observable world is the defining part of elective modernism and is worked out in chapter 2. As far as we know, this treatment is novel or almost novel.[23] Chapter 3 discusses the reach of the argument into spheres beyond science and concludes that it is minimal except when it comes to democracy; elective modernism has implications for the good society. For experts and for science to have their proper place in society, society has to be organized in a certain way. Furthermore, the values that hold science together have an overlap with the way a democratic society is held together. In chapter 3, a kind of political fable is used to indicate how such a choice might be put into effect in a democratic society, and a new approach to expert advice is invented to match the ideas. In chapters 4 and 5, the fable is put into scholarly context, with a manifesto summarizing our argument presented in chapter 6.

If the choice worked out in chapter 2 is to be convincing, it cannot be made to appear too easy. It is easy to defend science when it is a success – when it is predicting the movement of planets or the mass of subatomic particles, eradicating certain diseases or discovering the secrets of biological inheritance. But what has to be done, if elective modernism's *proposition* is to be as persuasive as it needs to be, is to show that science should be valued even at its times of epistemological and practical weakness. If science is to be a convincing moral choice, it must be loved and revered even when it is theoretically weak and practically inefficacious. Science has to be defended when it is failing to forecast the weather, continually going back on its treatments of medical ailments, proving impotent when it comes to predicting the economy, and proving helpless in the face of complexity in general – that is, when it doesn't work.[24] If science can be defended in these hard cases, then the easy cases are easy. At the end of chapter 2, econometric forecasting of the economy will be used as an example of the hard case. We will defend the preservation of econometric modelling for predicting national economic outcomes even though we know they have always been wrong and are likely to continue to be wrong, and even if we never make use of the results.

Our treatment of econometric modelling will give rise to a contradiction. One defence of it that will be mounted is that econometric modelling might get better if it is given a chance, whereas if it is not given a chance it will never improve. That argument is based on utility rather than moral worth, even though the preceding paragraphs have argued that arguments based on utility are dangerously weak. What is going on?

First, the argument from potential future utility is only one peripheral argument, while moral worth remains the central plank. But the deeper point is that the very notion of expertise contains the notion of utility. If no kind of expertise was ever efficacious, the meaning of the term 'expertise' would change and it would become something like 'opinion'. So, in spite of the major claim that the main defence of expertise has to be moral or commonsensical, we cannot avoid hoping that expertise is also efficacious and that part of the way of being of experts has to be to try to be useful; what we are saying is that they have only to try, not to succeed, to be doing good. But that the aim of succeeding is integral to the project of expertise should come as consolation to the rationalist.[25]

Part II
Elective Modernism

2

Choosing Science

Scientific values and the technical phase

Technological decision-making in the public domain involves the inter-mingling of expertise and democracy. What the distinction between technical and political phases allows – in fact, advocates – is that the two are kept separate as far as possible and we do everything we can to preserve the distinctive features of each. The differences between the technical and political phases were summarized in our 2002 paper in table 2.1.

The lower three rows of table 2.1 are relatively straightforward. Starting with row 2, the idea that participation is based on meritocratic principles follows directly from the idea that the technical phase is concerned with expert knowledge; even where concerns are raised by non-expert whistleblowers and other stakeholders, domain experts will need to evaluate the significance and impact of the claims.[26] The next two rows are closely related to this meritocratic requirement: first, experts will need to represent themselves in the technical debate. Expertise responds to unpredictable changing circumstances in real-time and is based on understandings which are often tacit, so no non-expert can be delegated to do

Table 2.1 Technical and Political Phases

		Phase	
		Political	**Technical**
Nature of	**Politics**	Extrinsic	Intrinsic
	Rights	Stakeholder	Meritocratic
	Representation	By survey	By action
	Delegation	By proxy	Impossible

the job. In contrast, because the political phase is assumed to be democratic, then anyone with a stake or their representative may participate and, as the vast literature on democratic theory attests, many different mechanisms exist for allowing this to take place.

The top row is more complicated and most of this chapter is about what we mean by 'intrinsic' politics. To understand the difference between Wave Two (mainly descriptive) and Wave Three (mainly prescriptive), it is important to distinguish between a descriptive 'is' and a prescriptive 'ought'. A well-known example in the STS literature – Shapin's study of the controversy surrounding phrenology in nineteenth-century Edinburgh[27] – makes the point. Shapin shows that the outcome, in which phrenology was effectively vanquished and proponents of the status quo retained their positions of influence and authority in the academic world, was closely linked to, and, in part, explained by, their connections and influence in the wider cultural and political life of Edinburgh society. This is the 'is' of the matter. The question is what follows from this description? One could argue that the scientific debate would have been concluded more quickly and efficiently if the influence of the various social and political factors had been brought immediately to centre-stage. But we argue that this would be incompatible with the idea of science. We argue that, despite the inevitable influence of local political factors in the technical debate about phrenology, the guiding principle should be to eliminate these effects so far as is possible. One conclusion that can too easily be drawn from the Second Wave is that, because science is affected by politics, one should forget the distinction – to act politically in a matter of science is to act scientifically. The

Shapin example reveals the flaw in this idea; would we want the fate of phrenology as a potential field of scientific knowledge to have been decided by local Edinburgh politics? The answer seems, a self-evident, 'no'!

The problem of demarcation

But what does it mean to act 'scientifically'? The fact that this question is a classic philosophical topic – 'the problem of demarcation' – makes it evident that science is hard to separate from other enterprises, at least in terms of a set of necessary and sufficient characteristics. If one cannot say what something is, then it is hard to choose it. Indeed, much of the power of Wave Two comes from exactly this point: showing that science is much like other activities means that the boundary between science and other kinds of social activities becomes blurred, and so identifying what it means to act 'scientifically' becomes even more difficult.

Luckily, science only *seems* hard to define. It seems hard because a mistake has been made about what it is to define something like science. Wittgenstein pointed out that, even though we regularly use the word and the idea of 'game' without being puzzled, we cannot define a 'game':

> Consider for example the proceedings that we call 'games'. I mean board-games, card-games, ball-games, Olympic games, and so on. What is common to them all? . . . if you look at them you will not see something that is common to all, but similarities, relationships, and a whole series of them at that.
>
> I can think of no better expression to characterize these similarities than 'family resemblances'; for the various resemblances between members of a family: build, features, colour of eyes, gait, temperament, etc. etc. overlap and criss-cross in the same way.[28]

'Science' is no more definable in terms of common properties than 'game' and *the problem* of demarcation arises only because it was thought that it should be. What we argue, following Wittgenstein, is that science can be readily recognized if it is thought of as characterized by a loose collection of family resemblances, mostly present but sometimes not.

Unfortunately, the notion of family resemblance needs to be used with some restraint if it is not to become meaningless. The problem is that anything can be linked to anything via a family resemblance with some intermediate object or property. Soccer uses a projectile – the ball. But rifle-shooting also uses a projectile so soccer and execution by firing squad are part of the same family and, since the stock of a gun is usually made out of wood, they also belong to the same family as carpentry and forestry – and so on. The notion of family resemblance does not work unless the family is restricted in some way in addition to overlap of properties.

Fortunately, the idea of a 'form of life', itself a distinctive feature of Wittgenstein's philosophy, provides the necessary restraint. Thought of sociologically, a 'form of life' is similar to many other sociological/historical concepts that capture the way life is typically lived in a social group.[29] The historical notion of 'epoch', Kuhn's notion of 'paradigm', the phenomenological notion of 'taken-for-granted reality' and, indeed, the concept of culture in general, all point to the fact that social groups both create their day-to-day worlds and operate boundaries around them. Science is a social grouping characterized by the typical actions and intentions of its members. In this, it is like other social 'collectivities', and its boundaries, membership and qualities can all be examined using standard sociological research methods.

To these ideas, we add that social groups should be thought as related in a 'fractal-like' manner. At the top end of the fractal are groups characterized only by the fact that they speak a common natural language, such as English.[30] At the bottom end are small groups of specialists who have narrow skills in common. Here the members share a more specialist vocabulary and way of talking that allows practitioners to understand and relate to each other and their world of specialized practices; we call this a 'practice language'. In between these two extremes are groups such as sports-players, soldiers, artists and scientists. These groups are of different sizes, they overlap and they are often embedded in one another but, whatever their size and whatever their substance, each group provides the set of 'formative intentions' of its members: it defines the ways in which they can legitimately intend to act *as members of that group*. Members of the Azande, *as members of the Azande*, can legitimately intend to divine witches by using the poison-oracle but they cannot legitimately intend to take out a mortgage; those who live in developed societies, *as members of developed societies*, can intend to take out mortgages but cannot intend to divine witches. Cricket players can intend to score a boundary but cannot intend, as cricket players, to kill members of the opposing side; soldiers can intend to kill members of the opposing side but, as soldiers, cannot intend to score a boundary. And so on. Members of any of these groups may also intend to eat or make love but those intentions do not help constitute the group of which they are members. Eating and making love are universal activities so in themselves they are not *formative* of social groups. *Culturally distinctive* ways of eating and making love do, of course, contribute to the constitution of forms of life.

Individuals usually belong to many overlapping forms of life. One can be both a cricket player and a soldier but the formative intentions are still different for each group and it is clear when an individual is acting as a soldier and when as a cricketer. To make the model work, it is important not to try to define the boundaries of groups by reference to which individual members belong to which group but to accept that *the group is the unit of analysis*.[31] It is then possible to say that when an individual is acting as a member of such-and-such a group then their formative intentions can include 'such-and-such' and cannot include 'such-and-such'. What the individual is at any one time is a set of formative intentions and formative actions belonging to one or more forms of life.

The vocabulary is a little clumsy: 'a form of life is built out of formative action types which are based on formative intentions'.[32] From here on, we will adopt the term 'formative aspirations' to refer to all of this. It implies that forms of life exist because their members aspire to act in certain ways most of the time, even if they cannot succeed all of the time. The term 'aspiration' ties the somewhat abstract notion of forms of life to individuals, but without imposing too rigid a straitjacket on their actions and behaviour.

Forms of life and social change

Two other features of this way of looking at the world are centrally important. Firstly, collectivities, or forms of life, change – but they change slowly. At one time, religion and science were close. One cannot foresee the future and it is possible that religion will one day become close to science again. What it is possible to say with complete certainty is that, today, religion and science are not close. Because groups change only slowly, if religion is, once more, to become close to science, it will not happen in the near future, and no decision about science that is made today should take into account that it is possible that disparate groups may merge in a remote future – after all, anything might happen in a remote future so to take the religion possibility seriously would imply taking every possibility seriously. For example, alchemy was once part of science and, conceivably, it could one day become part of science again, but this does not mean that today's science has to take alchemy into account.[33]

Secondly, because, in this way of thinking, the basic unit of analysis is the group, the individuals within groups may act in untypical ways without destroying the notion of what the group is. Thus, scientists may cheat, lie and act in self-consciously politically-biased ways without destroying the notion that the aspirations that are constitutive of science's form of life do not include cheating, lying and doing science in a self-consciously politically biased way. It might be that a scientist in the privacy of their laboratory manipulates some results to get to the truth of the matter as they see it, but that same scientist cannot announce to the world that their scientific results have been obtained by manipulation.[34] If they did announce such a thing, they would be challenging the formative aspirations of science, thereby challenging the very idea of science, and, as a result, ceasing to act as a scientist. Hypocrisy is often central to the maintenance of forms of life, or – to put a more positive gloss on the matter – a form of life is not destroyed if certain individuals act in ways that break the rules, as long as they do not celebrate the rule-breaking; if they do celebrate the rule-breaking, they are separating themselves from the form of life. It is the group aspirations and practice as a whole that give rise to the stability and identity of the form of life not the *action tokens* – the haphazard acts of individuals.

Problems of sociological and philosophical methods

The sociologist is faced with the problem of separating observed actions into two classes: those that do and those that do not constitute a form of life – the 'essential' and the 'accidental'. The way this problem is solved is through participation in the form of life in question which leads the investigator to 'understand' how it works (which is also the way 'behaviours' – in themselves meaningless movements, or 'strings' – can be properly assembled into meaningful actions).[35] That way, the action *tokens* that go together to constitute a formative action *type* can be separated from the action tokens that are not formative.

The 'problem of demarcation' arose because philosophers were looking for a logic of science and they continually found that any rule they invented to describe and bound science had exceptions. A sociological version of the problem of demarcation then arose because sociologists and historians looking closely at scientists' activities discovered that they did not – and, it was argued, could not – conform to any set of well-defined rules. But both philosophers and sociologists were making a mistake in thinking that for science to be special it had to match an idealized model. Forms of life do not match idealized models – they are full of exceptions. They are scattered with action tokens that do not match the formative action types. The important thing is the formative aspirations that characterize the collectivity, not the tangle of individual actions.

The difference can be seen in sociological or ethnographic practice in respect of science. Poor practitioners – those who do not know how to carry out an interview properly – are likely to elicit 'pat answers' from scientists – they elicit what the scientists think they ought to say rather than accounts of what the scientists do in their day-to-day practice. It was delightfully exciting when, in the 1970s, sociologists found out how to elicit the 'dirty details' of scientists' day-to-day practice, which differed so markedly from the existing formalized accounts of science. But bad interviews and 'pat' answers are not all loss: bad interviews elicit formative aspirations – what the scientists think they ought to be doing. The interview/ethnographic method is subject to two kinds of mistakes. The first mistake is when bad field researchers mistake an account of formative intentions for the day-to-day life of science. The second mistake is when good interviewers, discovering that science's day-to-day life does not match the idealized model, conclude that science is no different from ordinary life. But science is different from ordinary life and it is the aspirations of scientists – the guidelines that their actions often cannot match – that make it different, and the 'pat' answers capture some of what those aspirations are.

Philosophers, too, could have solved *their* problem of demarcation by relaxing the search for necessary and sufficient rules. They should have accepted that a practice like science is held together by intentions and aspirations which lead to a substantial change in a subset of actions; science is not held together by an unbreakable logic.[36]

Formative aspirations of science

What are the formative aspirations that go to make up science as an activity? There are a few insights from contemporary sociology of science but, mostly, elective modernism is old stuff. Elective modernism says the philosophers of science were right all along – pretty well all of them; they just didn't understand the social substance they were dealing with. It says the first wave of sociologists of science, before the sociology of scientific knowledge, were right all along; they just didn't understand the moral substance they were dealing with. The spirit of elective modernism is new but, mostly, the flesh is old.

This oldness of the flesh is very important. It is being claimed that science is characterized by a set of formative aspirations even if there are many accidental behaviours that do not match them.[37]

What gives these formative aspirations their stability and stops the accidental behaviours taking over? How do the aspirations survive in the face of the constant erosion by material self-interest? Why are there so few cheats and frauds in science? As we work through the values, it will become clear that they are all – or nearly all – based in a desire to reach the truth of the matter. They are robust because the worldview of the majority of scientists is the old-fashioned worldview that these values not only can, but will, reach the truth of the matter. It is the worldview of Wave One. Max Weber's 'protestant ethic' thesis does not posit that the accumulation of worldly goods *is* a sign of God's pleasure, only that capitalists believe it is a sign of God's pleasure. Likewise, scientists only have to believe that scientific values bring one closer to the truth; it does not have to be the case. If this is right, it is the old flesh of Wave One – what those brought up under Wave Two would think of as the touching naïveté of natural scientists – that keeps the values of the scientific community stable and leaves no doubt about what counts as integrity in the pursuit of scientific work. We should be desperate to preserve the moral imperatives that guided science under Wave One and this implies we are still in need of that simple model of science.

Popper's falsificationism

One well-known example can show how all this works. Karl Popper believed he had solved a major problem for philosophy of science when he invented falsificationism. The problem Popper thought he had solved was the problem of induction. It is the case that, however many positive instances of a rule such as 'all swans are white' are observed, one can never be sure that the rule is universal – a scientific law. But Popper said that a single observation – of, say, a black swan – could make one completely certain that the rule was false. From this, Popper developed a new *Logic of Scientific Discovery* (1959), based on the notion that the quintessence of science was the attempt to falsify scientific laws while holding those that had not been falsified as provisional truths. Popper's student, Imre Lakatos, then showed that Popper's

idea that there was an asymmetry between the process of corroboration and the process of falsification was incorrect. The single
observation of a black swan does not falsify the rule 'all swans are
white', because the black swan might be a white swan covered in
soot or paint so not really a proper black swan, after all. However
many black swans are seen, one must investigate each one in great
and fallible detail in order to establish that there is at least one
proper black swan. Thus, there is no asymmetry in respect of the
indefinite number of observations required, on the one hand, to
prove, and on the other hand, to falsify.[38]

From the point of view of elective modernism, what has just
been described is based on a series of mistakes. Firstly, while the
problem of induction/corroboration might present a problem for
philosophy of science, it does not present a problem for science
as a form of life. One of the formative aspirations of science is to
make many repeated observations of things in order to establish
general rules about those things. If scientists did not desire to make
repeated observations of things that could be repeatedly observed,
they would no longer be thinking as scientists. The fact that no
inductive generalization is completely secure does not affect the
formative aspiration.[39] The first mistake, then, is that there was no
problem for Popper to solve in the first place. For science, seen as a
social activity, there is no philosophical problem of induction, only
practical problems of observation! The second mistake is similar
to the first: there is no problem of falsification as adumbrated
by Lakatos. Scientists do try to find exceptions to their posited
rules and when they do find them they worry about whether a
rule has been falsified. Once more, the fact that no falsification is
completely secure does not affect the formative aspiration. Thus,
science is 'defined' by the fact that members of the group of scientists do aspire to corroborate and do aspire to falsify. But these are
not necessary or sufficient conditions of some act being the act of
a scientist – they are just things that have to be intended to happen
among the collectivity of scientists for it to be science that is going
on.[40]

To repeat, 'the trick' is to move away from the brittle,
exception-intolerant, *logic* of scientific discovery to the relaxed,
exception-tolerant, but still philosophically inspired, *sociology* of

science. This is based on family resemblance combined with the idea of forms of life. In the group of scientists thought about like this, corroboration and falsification are not alternatives fighting it out to solve a problem: on the contrary, there is no problem. In science as it is practised, corroboration and falsifications are both actions formative of scientific life, and collaborate to make science what it is.

Revealing what we have always known

We have always known the thing that has been said in the last couple of paragraphs. We have known about corroboration and falsification for more than half a century. Anyone trained in the field of social studies of science was aware of Lakatos's critique of falsificationism, but in our day-to-day academic conversations we continued to cite the unfalsifiability of some claim as a decisive criticism. And we were all happy with unrepeatability as a criticism too. This is what is meant by saying that this book is just trying to rearrange commonsense, or draw commonsense to the forefront of attention. It was the attempt to define a mathematics-like logic of science that made corroboration and falsification seem troubled and that began the process of deranging commonsense. It was the sociological critiques of science that grew out of too much respect for the contrast between scientists' actions and an ideal – the contrast between what scientists could actually accomplish and the ideals of perfection drawn from the philosophers – that deranged it still further. Commonsense had to be challenged by Wave Two if we were to understand how science is done; what we argue here is that if we are to continue to understand the world, then commonsense needs to be re-arranged. In mathematics, a single mistake destroys an entire proof; in the minds of too many philosophers, the cracks in the logic of science made the discovery of a definition of science recede just as it seemed within grasp; in the minds of too many sociologists of science, scientists' failures to follow the quasi-logical rules of science destroyed science as a distinctive activity. But in the kind of Wittgenstein-inspired sociology discussed here, descriptions of groups' ways of going on are not logical in form and are not destroyed by the discovery of actions that are less than ideal. Forms of life are more robust; they do not crack in the way that logic cracks. In the world of elective modernism, the jigsaw has been put together again and the pattern is pretty similar to what it was before the pieces were jumbled. It is the pieces that are different: in place of etched glass is hand-moulded clay, and the pieces don't fit tight, they more or less go together in an untidy kind of way.

The fundamental formative aspiration of science

There is one fundamental value to science that is so central and obvious that it is easy to miss. This is that the form of life of science is driven by the desire to find the truth of the matter, along with the belief that the truth of the matter can be found. The view associated with Wave Two is that the truth of the matter cannot be found, that there are only interpretations and perspectives and that it is naïve to steer one's desires and actions by the attempt to find the truth of the matter. It is the corrosive effect of this view that Wave Three tries to overcome.

That is all we have to say about the overarching value. The simple task that remains is to describe the smaller clay pieces – to build a list of the detailed formative aspirations that underlie the form of life of science. For each one, we can ask whether we value it more than the alternatives. If we do prefer the values to their alternatives, then we have shown to ourselves that we value science above other kinds of expertise when it comes to questions about the natural world. When we choose between experts, we will want to choose those who share science's formative aspirations and who try to do their work by reference to those aspirations. This is the heart of elective modernism. Much of the content is familiar – even mundane. It is the fact that we *choose* the values rather than try to *justify* them that is new.

Formative aspirations taken from traditional philosophy of science

Observation

The job has already been started. Logical positivism is the simplest philosophy underlying science; it bases meaningful synthetic propositions (those that provide new information about the world) on observation. Unfortunately, logical positivism did not stand up to close scrutiny because, among other things, it turns out that the only things that can be securely verified by observations are momentary sense atoms – such as 'green here now'. Anything more complex, such as 'this is a green wall', depends on building a generalization out of a mass of varied sensory atoms – for example, the sense of a green wall is the aggregate of many sense atoms of varying colours and intensities as well as depending on the idea of a wall – and the process of synthesis and the idea of wall are not purely matters of observation. So, verification through observation alone fails as a philosophical idea, but it does not fail as a sociological idea. We can still say that science as we know and understand it depends on observation as the basis of what is claimed about the world; the realization that observation is not the pure thing that it was hoped to be does not destroy observation as an aspiration and a practical guide for action.[41]

So here again is another prototypical example of the way our overall proposition is meant to work: if one wants to know about some feature of the world, does one prefer to listen to the opinion of one who has observed that feature of the world or one who has not observed that feature of the world? There are those who would prefer to listen to the opinions of those who have not observed it; they prefer the view of those who have seen the answer in a dream or divined the properties of that feature of the world with tea-leaves or entrails. But the elective modernist prefers to give more weight to the view of the person who has observed. And this remains the case even though we know that observation is inexact and impure and open to illusions, observer effects and the influence of the social group from within which the observation is being made. It remains the case even though we know that there will be occasions when the conclusions of the person who has observed will turn out to be less accurate than the view of the person who has not observed – the person with the tea-leaves. In spite of that possibility, the elective modernist prefers the view of the person who has observed. *And the elective modernist proper has to cleave to that preference without justifying it by reference to the successfulness of observation over other methods of knowing.* Not only is this *the way* of elective modernism, but it is also more robust than relying on success as a criterion: for all the reasons given for why observation is flawed and unreliable, there will be instances when observation turns out *not to be* more successful than other methods; but in these circumstances, elective modernism will be unaffected.[42]

There may be exceptions, but if these are at the level of the individual, not the group, there is no problem. Einstein is notorious for preferring his theory to any observations emerging from the 1919 solar eclipse that might conflict with them, since he was so certain the theory was right. But neither physics nor elective modernism is destroyed by that kind of exception; it is still the case that theory, or any other kind of *a priori* belief, cannot generally be preferred to observation in matters of observation. Or, if theory, or prophecy, or soothsaying is preferred to observation, the onus is on those who prefer it to explain why. Should it be that they believe that the whole world is an illusion, or that the word of the deity as located in old books is to be preferred to observation then, at least, the choice will be clear and it can be stated. There will be nothing left but a choice.

For elective modernism, then, logical positivism lives on as a basis of our aspirations for how to understand the observable world. And, of course, it always did. We have always preferred observations over non-observations where it was the observable that was at stake; we have just forgotten how to say it out loud.[43]

Corroboration

Now we try the same proposition on corroboration as an attempt to improve upon a single observation. We know that corroboration does not stand up to philosophical scrutiny – this is the problem of induction. It is the problem of induction that Popper was trying to circumvent with falsificationism. We also know it does not work because the practice has been examined in detail. For example, Collins, in his book, *Changing Order* (1985, 1992), showed that the replication of experiments was subject to the 'experimenter's regress' and could not, therefore, settle deep scientific controversies. And yet, as argued above, and as is argued in that early book, we still prefer to give more weight to the outcome of experiments which appear to have been successfully replicated over those which have not been successfully replicated. We would give more weight to the opinions of those who believe that it is proper for their findings to be replicated. Try imagining the world the other way round. Imagine a society in which the dominant view is that a single experiment is better than a repeated experiment and negative replications should always be ignored. In that case, anyone's first experiment, however superficial, would be the last experiment. Observations of the world would be fleeting and inconsistent. There would be no sense in which a physical fact or any other fact about the world would be thought of as stable or lasting. In that exercise of the imagination, we would have imagined a dystopia: if you prefer it to a world in which scientists try to confirm their findings through repeated observations, then there is nothing more to say.

The reader will be continually asked to imagine alternative worlds. It would be better still if the work of imagination was being done by novelists and science fiction writers. There should be a body of literature exploring the way worlds would look in which the formative aspirations of elective modernism did not hold. This is a new way in which science and the humanities could co-operate productively.

Falsification

The same proposition can be offered up to falsification. What is to be preferred – a world in which those who claim to have made observations are willing to set out the conditions under which they could be

shown to be wrong and actively invite the test, or a world in which
they would consider this unnecessary or inappropriate? The latter
would, once more, be a world in which we would not know how
to change or question an observation. All observations would be
revelations. Once more, it is the imagination that is needed to show
how central to the world as we know it the idea of falsification is.
And, to repeat, it is central even though it is itself imperfect.[44]

Formative aspirations from Mertonian sociology of science

We now turn to those formative aspirations identified by sociologists of science who aimed to discover what made science special.
It is well known that Robert Merton described a set of norms
that he believed constituted science. It is equally well known that
Merton's norms do not work as a description of everything that
scientists do in the name of science, and that 'counter-norms' were
later identified that appeared to undermine the whole project.[45]
Nevertheless, we want to argue that Merton did reveal something
important about the nature of scientific work.

Effectively, Merton's error was his desire to get an 'ought' from
an 'is'. He took it to be the case – something pretty well unquestionable at the time – that science was the most effective way of
gaining knowledge. From this, he argued that the norms of science
should be adhered to because they were efficacious:

> The institutional goal of science is the extension of certified knowl-
> edge. The technical methods employed toward this end provide the
> relevant definition of knowledge: empirically confirmed and logically
> consistent statements of regularities (which are, in effect, predictions).
> The institutional imperatives (mores) derive from the goals and the
> methods. The entire structure of technical and moral norms imple-
> ments the final objective. The technical norm of empirical evidence,
> adequate and reliable, is a prerequisite for sustained true prediction; the
> technical norm of logical consistency, a prerequisite for systematic and
> valid prediction.[46]

Merton, however, has a tendency to cover all bases and he also said, immediately following this remark:

> The mores of science possess a methodological rationale but they are binding, not only because they are procedurally efficient, but because they are believed right and good. They are moral as well as technical prescriptions. (p. 270)[47]

It is almost certain that Merton thought the driving force for the norms was scientific efficacy, but he was probably saying that scientists internalized them as more than this – as good norms of action. In this book, the rationale is different. We use only the idea that the norms are good in themselves; we do not claim that they are efficacious because, with the huge enriching of our understanding of science since the 1970s, we do not think this idea should be taken as the central justification for science. On the other hand, we do think that the model of science used by scientists, and at least some social scientists, does lead straight to the norms identified by Merton, and it is scientists' idea of science that drives science as an institution and gives it the potential to provide moral leadership for society as a whole.

Merton in historical context

Merton initially wrote his essay in 1942 and it is clear that he set science against European fascism. For example, under the heading of the norm of universalism, he wrote:

> The circumstance that scientifically verified formulations refer in that specific sense to objective sequences and correlations militates against all efforts to impose particularistic criteria of validity. The Haber process cannot be invalidated by a Nuremberg decree nor can an Anglophobe repeal the law of gravitation. The chauvinist may expunge the names of alien scientists from historical textbooks but their formulations remain indispensable to science and technology. However echt-deutsch or hundred-percent American the final increment, some aliens are accessories before the fact of every new scientific

advance. The imperative of universalism is rooted deep in the impersonal character of science. (p. 270)

In later editions, he added a footnote contrasting the norm of universalism with Russian communism:

in an editorial, 'Against the Bourgeois Ideology of Cosmopolitanism', *Voprosy filosofii*, no. 2 (1948), as translated in the Current Digest of the Soviet Press 1, no. 1 (1 February 1949): 9: 'Only a cosmopolitan without a homeland, profoundly insensible to the actual fortunes of science, could deny with contemptuous indifference the existence of the many-hued national forms in which science lives and develops. In place of the actual history of science and the concrete paths of its development, the cosmopolitan substitutes fabricated concepts of a kind of supernational, classless science, deprived, as it were, of all the wealth of national coloration, deprived of the living brilliance and specific character of a people's creative work, and transformed into a sort of disembodied spirit . . . Marxism-Leninism shatters into bits the cosmopolitan fictions concerning supraclass, non-national, 'universal' science, and definitely proves that science, like all culture in modern society, is national in form and class in content.' This view confuses two distinct issues: first, the cultural context in any given nation or society may predispose scientists to focus on certain problems, to be sensitive to some and not other problems on the frontiers of science. This has long since been observed. But this is basically different from the second issue: the criteria of validity of claims to scientific knowledge are not matters of national taste and culture. Sooner or later, competing claims to validity are settled by universalistic criteria. (1979: 271)

It is clear that Merton thought that democracy best embodied scientific values so democratic societies would best foster an efficacious science. Responding to, among other things, the notorious theories of the superiority of Aryan science promulgated by Johannes Stark, published in 1938 in *Nature*, Merton wrote:

However inadequately it may be put into practice, the ethos of democracy includes universalism as a dominant guiding principle.

Democratization is tantamount to the progressive elimination of restraints upon the exercise and development of socially valued capacities. Impersonal criteria of accomplishment and not fixation of status characterize the open democratic society. (p. 273)

Lest there can be any doubt, Merton concludes his article:

Conflict becomes accentuated whenever science extends its research to new areas toward which there are institutionalized attitudes or whenever other institutions extend their control over science. In modern totalitarian society, anti-rationalism and the centralization of institutional control both serve to limit the scope provided for scientific activity. (p. 278)

Given the world as it was then, this is probably what almost anyone who was not a fascist or communist would have argued.

Merton's norms of science

Merton's four norms, to which we will add extensively, have been given the acronym 'CUDOS'. Its elements were:

Communism, which entails that scientific results are the common property of the entire scientific community.

Universalism, which means that all scientists can contribute to science regardless of race, nationality, culture or gender.

Disinterestedness, according to which scientists should not present their results entangled with their personal beliefs or activism for a cause. Scientists should have an arm's-length attitude towards their findings.

Organised Scepticism, which requires scientific claims to be exposed to critical scrutiny before being accepted.

Because Merton was a sociologist, rather than a philosopher, his scheme does not crumble at the first discovery of a logical imperfection, but is vulnerable should it be found that scientists do not act according to the norms yet still regularly produce good science.[48]

And this seems to be the case – there are many instances of successful science that do not match one or more of the norms – for example, the Manhattan Project, the Second World War research project that created the first atomic weapons, was a long way from following the norms of universalism and communism.[49] Thus, the 'is' is not what he thought it was, so there is no justification for the 'ought'. But, under elective modernism, Merton would have had no need to try to justify the 'ought'. Under elective modernism, Merton's norms are just good in themselves in a self-evident kind of way.

Communism

Communism is the idea that scientific knowledge should be shared so that all scientists have equal access to it and collective responsibility for its development and scrutiny. It is, perhaps, the hardest norm to justify as a general good for science but, to the extent that organized scepticism depends upon it, it would seem to be important. To use the same style of argument we will use for the other values, we will ask if we can see why a society in which scientific work was based around a norm of communism would be preferable to one that took its opposite – secrecy – as its default option.

A norm of communism implies a society in which a request to a scientist for information or advice should normally receive a positive, or at least helpful, response. This seems self-evidently preferable to one in which scientists are under no obligation to help each other and do not even have to provide a reason for refusing to co-operate. A science organized around a norm of secrecy is also one in which lying and deception are endemic. It is hard to see how such a system would be preferable to one in which the sharing of knowledge and information is regarded as the right thing to do.[50]

Universalism

Compared to communism, justifying 'universalism' is easy. To choose universalism, one need only ask if, in general, one wants to live in a society in which a person's opinion in respect of the observable world is weighted according to their race, nationality, culture or gender? The answer seems to be a self-evident 'no'. At the very least, the onus is on those who would prefer a different choice to justify it. One might argue, for example, that the world of natural science has been weighted in favour of men for so long that it is time to redress the balance by deliberate weighting in favour of women. But this is not to argue for the weighting to be always a matter of gender, only for a temporary exception in order that the value of universalism be met more successfully.

Disinterestedness

Turning to disinterestedness, can one prefer to live in a society in which views about the natural world are weighted according to personal beliefs or activism in a political cause? It seems, again, that the answer is a self-evident 'no' and it is only a 'yes' that requires justification, for, at best, the occasional exception. Were it not so science would be a continuation of politics, because every so-called 'scientific finding' would be designed to fit a political agenda – it would be Lysenkoism and its equivalents every time and everywhere. This would surely be a dystopia.[51] Note, once more, that much contemporary science studies-inspired analysis argues that every scientific conclusion *is* invested with politics, but elective modernism argues that this does not mean that we must *aspire* to make every scientific act a political act. Rather, we should be disappointed that we cannot find the pure realm of science for which we must continue to strive.[52]

Organized scepticism

Can one prefer to live in a society in which scepticism about claims about the physical and biological world is discouraged rather than encouraged, so that claims can be made capriciously without risk? Again, it seems that the answer is a self-evident 'no' for just the same reasons that replicability and falsificationism were pre-ferred, and it is hard to think of any exceptions that might need justification. Again, the suppression of criticism, especially criticism of authority, seems to be a characteristic of dystopias.

Additional formative aspirations

There are a number of additional aspirations and values that can be said to be formative of science which can be derived from critical reflection on science as a form of life. It is difficult to see how anyone could argue that violating the following norms would improve science in any way; some are excellent values for living in general, but more 'integral' to science.

Honesty and integrity

Because, under elective modernism, we are talking of formative aspirations, we can include things as basic as honesty and integrity, both of which are missing from the Mertonian scheme.[53] Of course, honesty and integrity are such universal values, necessary for the functioning of society in general, that it might be argued that they are not formative of science *per se*. But integrity in the search for evidence, and honesty in declaring one's results, are *integral* to science, not just important values. Where evidence is fabricated or deliberately misreported, then science simply is not being done, whereas other forms of life are not negated by dishonesty in quite the same way. This is not to say that all those who call themselves scientists should always be honest; there may be higher moral imperatives. Ludwik Fleck, confined to a concentration camp, produced ineffective typhus vaccine for the German army. But Fleck was not doing science at that point, he was waging war.[54]

Locus of interpretation

A corollary of the preference for experts over non-experts and for observation over non-observation as the route to knowledge about the natural world is that, in the case of science, the 'locus of legitimate interpretation' must be close to the producers of knowledge. The locus of legitimate interpretation (LLI) is where legitimate critics and interpreters are found.[55] Compare science with the arts, particularly the adventurous arts. Art is meant to be consumed and the LLI is with the consumer – or the consumer's proxy, the critic. A formative aspiration of scientists, on the other hand, has to be to have their work valued by the most expert of colleagues – the public comes very far down the line in importance. The modern stress on public understanding of science is a matter of securing financial and institutional support, not a matter of technical evaluation. The value of art, in contrast, is entirely a matter of the evaluation of critics and the public. When rejected scientists turn to the public for affirmation of their findings, they are no longer acting as scientists and are quite properly distrusted by the body of scientists. Now imagine a society in which public acclaim for scientific findings was valued more than expert acclaim – again, it would be a dystopia.[56]

Clarity

Artists themselves have no role in the interpretation of art, while scientists are the only consumers with sufficient observational skill to provide science's organized scepticism. The interpretation of art can be manifold and the artist's intention can be to provoke a multitude of interpretations – sometimes, the more interpretations the better. The scientist, on the other hand, must aspire to convey only one possible interpretation – the correct interpretation. What follows from this is that clarity is an imperative in science in a way that it is not in other cultural endeavours. Where multiple interpretations are encouraged, obscurity can be a virtue but where the aspiration is to produce only one interpretation, clarity is to be preferred. Obscurity is also privacy, and privacy is an obstacle to organized scepticism. It follows, from the fact that observation and scepticism are virtues in the pursuit of knowledge about the natural world, that clarity is also a virtue.[57] There is an irony here, since science should seek maximum clarity and accessibility even while acknowledging that the only proper assessors of a scientific claim are small elites, while artists often seek obscurity or multiple interpretations even while accepting that the proper assessors are the public.

But clarity is a virtue in itself. Consider which kind of society is preferable: one that encourages those who provide knowledge about the observable world to be obscure, or one that encourages them to be clear in their claims. It seems self-evident that the society that encourages clarity is the virtuous one, while one that encourages obscurity and mystery approaches the dystopian end of the spectrum.

Individualism

Another feature of science as we know it is that, in principle, it must be supposed that an individual can enter into dialogue with nature and hear more clearly than the rest. This, again, follows from the fact that observation is the key and that some people are in a better position to observe, or are better observers, than others. The scientist must, in principle, be the person who is ready to recognize that the emperor has no clothes or that the Earth goes round the Sun, while all others concede to the emperor's view of things or believe that the Sun goes round the Earth. What, in 1979, Thomas Kuhn called 'The Essential Tension' captures the fact that, in science, individuals feel they have the right to stand up against the powerful majority, but science is, nevertheless, held together by the powerful majority. Once more, consider the alternative: imagine a society in which individuals' claims about observable matter were never considered if they clashed with the view of the majority. Once more, it would be a dystopia. As Kuhn intimates, it would also be a dystopia if every individual scientist's views were taken equally seriously.[58]

Continuity

Science is continuous, not discontinuous, within domains. This is not the old 'unity of science' thesis that implied that scientific method was the same everywhere. Methods vary according to topic, and it will be explained below that social science uses very different methods from natural science. There is, however, unity across sciences in respect of its formative aspirations and the characteristics of its form of life as outlined here. Even this, however, is not what is meant by 'continuity' as one of the formative aspirations. What is meant is that, where originality is claimed within a domain of science, it is not the break with existing science that is valued. The aim of a scientist who discovers something new and radically different will be to persuade others in the scientific community to understand it and take it up within the existing community. In spite of the fact that we know there are radical discontinuities in science – with Thomas Kuhn's paradigms being the most well-known expression of this fact – scientists, *qua* scientists, strive to minimize the change required for new ideas to be accepted: they want to preserve as many of the existing institutions as possible and have their ideas acclaimed by the existing scientific community, rather than create a new community. New communities are created only reluctantly and in the face of determined resistance from the old. This is quite different from the aim of political revolutionaries, for whom radical change is the essence of what they are trying to accomplish. Whose view does one prefer when it comes to observing the world – one who claims that their radically novel observations nevertheless fit with science and can be shown to be valid by science's established procedures, or one who claims that they can only be understood if science's procedures and most of its existing knowledge are overturned? The answer, we believe, is clear. A society characterized by continual revolution in the matter of the relevant affirming population would have no stability.[59]

Open-endedness

Science is never over. All scientific claims are falsifiable and might be falsified. The work of science goes on for ever. To whose opinion does one give preferential weight? One who says that their claims represent the end of science, or one who says this is the best they can do for the time being? Once more, we believe the answer is obvious.[60]

Generality

Scientists value generality: the wider the application of some claim, the easier it is to verify or falsify. Thus, in social science, a case study which claims that no characteristic of the case is generalizable to other cases – that the duty of the scientists has been discharged simply by a rich and detailed description of the particular case – is not testable by anyone who has not done the case study. To make the work open to criticism, it is much better to describe characteristics of the case that ought to apply to similar cases. In general, the broader the application of the claim, the more scientifically worthy it is.

Expertise

A foundational feature of science is, of course, the high value that is given to expertise. Scientists aspire to be as expert in their specialism as possible and to give special weight to the views of experts. The aspiration is to value experts for what they can accomplish rather than for their position of authority even though separating the two is difficult in practice. This valuation, of course, is related to science's locus of legitimate interpretation.

Science as a logical machine and as a form of life

Table 2.2 is a summary of much of what has been said, with the first four lines marking some more general contrasts between science conceived of as a logical machine and conceived of as a form of life. Starting at the top, earlier philosophies and sociologies of science held that, if there was to be a distinctive thing called science, it must have deep and secure foundations; elective modernism says

Table 2.2 Science under Elective Modernsm

	LOGIC of SCIENCE	SCIENCE AS FORM OF LIFE
1	Deep-rooted justification	Shallow-rooted
2	Truth	Expertise
3	Facts	Approach and norms
4	Right decisions	Best decisions
5	Observation	*Observation*
6	Corroboration/replicability	*Corroboration/replicability*
7	Falsification	*Falsification*
8	Universalism	*Universalism*
9	Disinterestedness	*Disinterestedness*
10	Openness to criticism	*Openness to criticism*
11		Honesty and integrity
12		Locus of Legitimate Interpretation
13		Clarity
14		Individualism
15		Continuity
16		Open-endedness
17		Generality
18		Value of expertise

that a robust defence of science will have shallow foundations — if moral choices and a preference for experts are thought of as shallow. Traditional philosophy of science is too ready to make true facts/findings the aim of science, and the consensus about true facts the indicator of whether a piece of science has been correctly executed; elective modernism takes it that approaches, not facts and findings, are the key to science; the way things are done is more central to our society than what is found out. Where science and technology feed into decision-making in the public domain, traditional philosophy sought the *right* decisions; elective modernism recognizes that the speed of politics is too fast to wait for science and the aim is to make the *best* decisions that can be made even if they turn out to be wrong in the long term.

The italicized entries starting at row 5 in table 2.2 show where elements from the logic of science have been transmuted into the softer elements of a form of life. It is Wave Two of science studies that is responsible for the transmutation. The final few rows of the Table show the new elements that have been added under elective modernism and Wave Three of science studies.

The sensation that we hope to engender in at least a subset of our readers is the same as the sensation that elective modernism engenders in the authors: a sense of relief. We have always known that the formative aspirations of science are good things to live by but the work of the last half-century of science studies has made it impossible to say *why* it is so and, for many, embarrassing even to say *that* it is so. The problem arose out of having to choose between an idealized, logic-like and totally unforgiving model of science and the richer description that emerged from around the 1960s onward. Even those immersed in the second tradition began to see its danger where difficult questions involving medicine, global warming and the like surfaced.[61] Unfortunately, the only alternative seemed to be the old, unreflective, admiration of science. Under elective modernism, we can relax: the formative aspirations are good things even when they cannot be fully realized in practice. Indeed, as social practices, they can *never* be implemented in this way! Now we can endorse them without sacrificing the rich critiques of the philosophical rules and the rich sociology of scientific knowledge. The formative intentions are just good in themselves and they remain good in themselves even when that old, idealized, unforgiving, model of science is set aside.

The hard case:
defending science when it is ineffectual

If we are sincere about defending science on the basis of its moral virtues, then we must demonstrate that we can defend it even when it has no utility. Suppose one is to make a decision and one is looking for advice: should one prefer the advice of those who know what they are talking about or the advice of those who do

not know what they are talking about? The answer to that question seems self-evident – take the advice of those who know what
they are talking about! But it is a little more complicated. The
expression, 'they know what they are talking about', is often used
in a loose way to mean that 'they' are likely to be right in their
judgements. Here, however, we are using the expression literally.
A person who knows what they are talking about has studied, or
spent much time getting to know, the thing in question. Such a
person 'knows' in the sense of having long and intimate acquaintance with the thing – they are familiar with the thing in the way
that they are familiar with people that they know. They are the
people who have made the observations in matters where the
observable is at stake. This, however, does not necessarily mean
that their judgements will be right. A husband knows his wife and
yet still may be surprised to find she has a lover; a driver may have
long experience of the roads and still make a poor judgement that
leads to an accident; a property speculator may long be familiar
with rising house prices and still be surprised to find they are in
negative equity. Knowing what you are talking about does not
make you right, it makes you familiar with your subject. If familiarity equated with rightness, then experts would never disagree.

We begin with our standard move. What would a society be
like in which all expertise has been levelled out and anyone's
opinion is considered as good as anyone else's on any matter?
Such a society would be a dystopia. Imagine it: when you want to
know something, you pick on anyone to ask, and it does not occur
to you that there are people with more or less valuable opinions
on the matter. To choose a person to ask about something, one
could use a random number table; there would, of course, be no
encyclopedias, and perhaps no books, because there would be no
concept of a source of information. That is the first proposition.
The second proposition involves its application to an example, the
insecure science of econometrics.

Econometric forecasting of national inflation rates, unemployment rates and so forth has been shown – in, for example, Evans's
Macroeconomic Forecasting – to be reliably inaccurate.[62] For short,
we'll refer to this kind of forecasting simply as 'econometrics',
though, of course, the subject has other applications, many of

which are not so unreliable. Given the usage adopted here, we can say that econometricians are reliably wrong. But what would the world look like if we did not value the advice of econometricians more highly than that of 'the person in the street'? Imagine we are having a university seminar on the prediction of inflation and unemployment rates, or a decision-making meeting that turns on predicting economic growth over the next year. Who should we invite to talk to us? Econometricians are generally well-paid and we are likely to have to cover their travel expenses so why not just invite the first person who passes the door and listen to what they have to say? Firstly, it is possible to see, we believe, that even though the person in the street is likely to be no less wrong in predicting rates than the econometrician, to give up the institution of the learned seminar is to begin the dissolution of our society. A society in which the weight of an opinion is not increased according to the expertise of the opinion holder – the extent to which they know what they are talking about – is a society that would have quite different institutions and procedures from those of the developed and developing world. To listen harder to an econometrician talking about next year's inflation rate than to any person in the street is, therefore, a good thing even if the econometrician is wrong – as he or she is likely to be; this is because to listen is to recognize the importance of scientific values and preserve the institution of expertise in general.[63]

Secondly, by continuing to listen to econometricians more than others in the matter of next year's rates, we are cherishing them in particular and making it more likely that they will survive as a profession. That increases the chances that at some time they will become less unreliable. If we wipe them out, then there will never be a successful econometrics.[64] This, of course, is not a foundational argument for elective modernism since elective modernism rejects scientific efficacy as a criterion, but it might be a good criterion for those who want two strings to their bow. Of course, there might be competing groups of experts who do better at predicting the financial and unemployment rates, or who could improve the econometricians' ability to forecast – analysts of wider aspects of human behaviour are candidates for this role – but that is just to point to one of the ways in which econometrics might improve itself as a science. It is also the case that sometimes we have to abandon bodies of experts because we come to believe they will never succeed in satisfying us in respect of their abilities; for example, alchemists have gone that way. The point is only that, in nearly all such cases, abandoned groups should be replaced by other groups of experts, rather than persons in the street.

A difficulty for the argument is that there are many competing groups who know what they are talking about when it comes to making predictions, including economic predictions. Astrologers are an example. Astrologers are experts. In the sense of the phrase as it is used here, astrologers know what they are talking about. An economic astrologer (assuming there was such a thing) would know the subject of predicting economic outcomes through studying the positions of the planets and constellations. How does one choose econometricians to predict economic outcomes over these other groups? Even if one is not going to pick just anyone off the street, why not choose an astrologer over an econometrician to predict next year's inflation rate? Astrologers are experts and, given that we are talking of today's econometricians predicting inflation and unemployment rates, we can say that the astrologers have as good a record of success and are likely to be no worse at making predictions than the economic modellers. The answer – the reason we pick the econometricians over the astrologers – is that we prefer scientific experts because scientific experts aspire to fulfil the characteristics of science listed in table 2.2, line 5 onwards. We believe that econometricians will adhere to universalism rather than particularism; to disinterestedness rather than the pursuit of personal interests; they will endorse organized scepticism rather than protect themselves from criticism; they will trust observation above revelation; they will be willing to state how a claim could be falsified rather than present their findings as absolutes; they will value the criticisms of other econometricians more than the acclaim of the public; they will understand that clarity is better than obscurity; they will allow a space for individuals to stand outside the mainstream of their discipline even while they try to maintain collective consensus; they will try to make their work continuous with the existing body of work in their discipline; they will allow that their science is open-ended; and they value expertise.

Unfortunately, not all econometricians share all these aspirations. Furthermore, there are some astrologers who share some of them. But if we want advice from people who adhere to these values, then we are much more likely to get what we want from econometricians than from astrologers and, should empirical studies find that not to be so, then econometrics needs fixing. Our argument, remember, is prescriptive as much as descriptive.

We should, of course, avoid individual econometricians who are known not to adhere to the values. To repeat, the fundamental reason for the choice between econometricians and astrologers under elective modernism turns, in the first place, on the values of the people who make up the groups of experts, not their likelihood of success. And it cannot be that the values are justified by the success they bring because, in today's econometrics, there is no success. That is why it is a robust choice, even though its foundations are shallow, not deep. Only under elective modernism can the choice of econometricians over astrologers be maintained in spite of the fact that astrologers are just as likely as econometricians to get next year's inflation rate right.

Interim conclusion

So far, we have tried to accomplish two things. The first is to flesh out in more detail what we mean by the 'intrinsic politics' of the technical phase. Intrinsic politics means that experts try to resolve their differences whilst adhering to the values we have just described. This means disputes are resolved by reference to observation and theory and without resorting to arguments about the consequences or desirability of particular outcomes; to do this would be to make the political concerns extrinsic and to stop acting 'scientifically'.

The second, and by far the more important, is to demonstrate that these values are good values. It should now be close to self-evident that, where the observable is concerned, we should all prefer the views of scientific experts and those who share their values over the views of all other kinds of experts, and over the views of non-experts. We should prefer the views of those who know what they are talking about, and we should prefer the views of those who have come to know what they are talking about while following the formative aspirations of science. This preference rests upon the dystopian quality of any other preference. That is what is meant by elective modernism.

3

Elective Modernism, Democracy and Science

Elective modernism is based on moral choices and the attempt has been made to show that they are good choices. In this chapter, we explore its implications for the relationship of science and society by setting out a model for the 'political phase' of technological decision-making in the public domain. To do this, we look first at the 'reach' of elective modernism – what kinds of cultural enterprise does it bear upon? We argue that its reach is quite limited. Then we argue that, though limited in its cultural reach, elective modernism is less limited in its political reach, having something to say about the nature of democratic societies and something to offer them. To make this final point, we invent a new kind of statutory institution for dealing with the relationship of science and society – an institution that reflects these ideas and exemplifies how the technical and political phases can work together. This institution is called 'The Owls'.

Elective modernism's reach

While elective modernism champions science in the domain of the observable, it is not a fundamentalist ideology. Its reach is the observable world. It has, for example, nothing to say about aesthetics; in art, the ways of going on are different and the ways of judging are different. It has nothing to say about religion except where religion makes claims that bear on what can be observed. Thus, science and religion clash over creation and evolution and here the elective modernist will choose science over religion because the tenets of elective modernism are incompatible with even that most subtle version of creationism, intelligent design. Intelligent design fails to live up to at least four of the formative aspirations of science identified in the previous chapter: it cannot be falsified; it is not open-ended as the Deity is the ultimate answer; it is rooted in obscure books and revelation rather than observation; and, by invoking the agency of an omnipotent creator, it is not continuous with existing scientific explanations that rely on natural causes. Given this, whose opinions in the matter of evolution should be given most weight? Should it be to those who aspire to act according to the ideas of falsifiability, observation and open-endedness, or those who do not? If one prefers science, the answer is obvious. Only if one prefers non-science as a way of understanding what we can observe of the natural world would intelligent design be the better choice.

But elective modernism is not a new religion; it does not demand that all our thinking, even about the universe, come under the aegis of science. That the laws of science and the universe itself might have been formed by a creator is not incompatible with elective modernism; that humans have a soul that goes to heaven after death is not incompatible with elective modernism; that the creator is benign is not incompatible with elective modernism. Elective modernism simply has nothing to say about these things because they are not matters of observation. Elective modernism says that scientists, as scientists, should not try to prove or disprove things to do with spirituality (or art and the like) except where such things have repeatable, observable consequences. In this sense, elective modernism is far from logical positivism and the many other kinds of positivism that equate the non-observable with the meaningless.[65] Should observable consequences for any of these things be found, then they would become the business of elective modernism and subject to the same preference for scientific investigation.

Thus, elective modernism has no objection to those who try to make matters of spirituality observable. If someone wants to test for the existence of life-after-death by waiting for the newly deceased to transmit a message given to them before death in a sealed envelope under controlled conditions, or test for out-of-the-body experiences by reading the code on the top of a wardrobe while they rest in bed, or test for precognition by guessing cards, or test for psychokinesis by trying to make a random number generator less random, why shouldn't they? These are hard experiments which experience tells us are almost certainly doomed to fail, and still more certainly doomed not to convince anyone even if they produce a seemingly positive result, but something scientific would be going on. To say that no scientist should try such things would imply that the limits of the observable world are already set, and that all the kinds of explanation are already known, and that the constituents of the universe have been counted for the final time. All this would be in violation of the formative aspiration of open-endedness. What elective modernism has to say is that, if these things are to be investigated, they should be investigated according to the formative aspirations of science. There are hucksters and charlatans among both the proponents and sceptics of the paranormal; they make up a penumbra that surrounds the few innocents at the heart of the scientific research. The hucksters and charlatans are not to be treated with respect; the innocents are to be treated with respect, however hopeless their project.

That elective modernism has no objection to research on the paranormal does not mean it actively encourages it; it is not the job of elective modernism to encourage or discourage any kind of scientific research. Rather, the aim of elective modernism is to provide some guidance on how scientific and other expert advice should be weighed and used when technological decisions are being made. The argument to come in the remainder of this book is that experiments in the paranormal and the like should be ignored as far as practical matters of policy are concerned because they are well past their 'sell-by date'; the paranormal has been given a very long run for its money without demonstrating good control over its subject matter and has fallen off the list of things that policy-makers should consider seriously.[66]

The policy imperative is different from the scientific imperative, and so the political phase must be different from the technical phase: science and politics, though linked, are not inter-changeable. The policy imperative is far narrower and has no commitment to open-endedness because decisions have to be made relatively quickly, and decisions are, by their nature, closed once made. But closure for policy is no reason to impose closure on science. All it means is that a science that is sufficiently settled to be taken-for-granted by policy-makers is no longer a priority when public funding is allocated; others may have different priorities and choose to fund such research privately and this is entirely their choice. There is no reason for scientists to attack those who have new ideas about how to make weird things observable or who continue to chase what almost everyone else believes to be a lost cause. When they are attacked by scientists, it is usually because ordinary science is mistakenly thought to rest on a set of quasi-religious beliefs with *unobservable* consequences. Such beliefs include that science has the potential to explain all phenomena or that the set of all possible explanations is already in place. It seems to be thought by those who mount such attacks that the survival of science depends on the extinction of competing beliefs – just as in the case of competing religions. This is a long way from elective modernism. There is no reason why minorities and mavericks should not continue to exist; in fact, it is probably a sign of a healthy science that they do exist. All that mainstream scientists and policy-makers need to do is ignore them: they do not have to attack them. Those who feel the need to attack think they are science's strongest supporters but they are not adhering to the values of science, they are betraying science. Of course, mavericks who mislead the public in a significant way, such as those involved in medical controversies who prey on fears of hypothetical risks, do need to be countered, but that is a policy matter to be resolved through better communication, not a scientific matter that needs more data. None of this touches on the question of scientists without integrity, who need to be countered by every means.

The social sciences

Are the social sciences proper sciences? Yes – at least they might be if conducted in a certain way. It is often thought that certain methods define the sciences, but this is not so. There is no unique method – only a plurality of scientific *methods*. There are, however, the formative aspirations of science listed in the previous chapter, and when social scientists adhere to them they constitute a science. That said, the methods of the natural and social sciences are importantly different. It is not the case that mathematical analysis is the essence of science; it is not even the case that mathematics is the essence of physics. It is not even the case that scientific methods must be 'objective'. All that is required is that observations be potentially repeatable, falsifiable, and so forth. Thus, a perfectly respectable method of analysis in sociology is 'participant comprehension'.[67] The aim is comprehension of the tacit ways of going on of groups of humans – their forms of life. This is a 'subjective' method, unavailable to any natural scientist because one cannot comprehend atoms or cells in the same way. But, of course, even in physics, all observations are subjective in the first instance – someone has to use their brain and body to read the meter. It is just that, in a physics experiment, the data that are to be read are first inscribed on metal and plastic; in the case of a social science, the data to be read are inscribed on the collective brains and bodies of social beings, and that requires a special kind of 'meter-reading'. The results of this doubly subjective method are, nevertheless, generalizations about society and, as long as the results are described in a sufficiently general and clear form, they can be tested by others using the same method; if others do not find the same results, then something has gone wrong.[68] It is, then, science that is going on, even though the initial observations turn on the subjective. To repeat, the natural and the social sciences are divided quite radically by methods but, in principle, they are not divided by their values or by their credentials as sciences. We do not say this is a description of all the social sciences since many who call themselves social scientists, particularly sociologists, conduct their research and scholarship as though it belongs to the humanities rather than the sciences, and there the values can be different; we say it is a claim about principle covering those social scientists who wish to *proclaim* themselves to be scientists in a self-conscious way. The social scientists who do their work more in the tradition of the humanities than the sciences are simply choosing to work in a different way.[69] So science, as it has been described here, reaches into the social sciences, and so does elective modernism.

Another feature of the sociological approach to social science taken here may help to illustrate further what elective modernism is and is not. Social life is deeply mysterious – at least in respect of our current knowledge. Forms of life seem to be fully describable, but not to be further analysable. Forms of life seem to be the basic constituent of the social universe – the fundamental unit of analysis. There do not seem to be any sub-form-of-life particles in the way that there are sub-atomic particles. The mistake is to think human individuals are sub-form-of-life particles. In contrast, under the view being outlined here, individuals are made up of the forms of life they participate in, not the other way round. Forms of life are sub-human particles, as it were. 'Collective tacit knowledge', the tacit constituent of a form of life, seems impenetrable – at least for the time being.[70]

Certain kinds of scientific fundamentalists would consider such a view to be almost sinful. They would insist that all human action and behaviour must be explicable in terms of evolution and the biological and neurosciences. Even though the work has not been completed, they would insist that we already know this and that every scientist must endorse such a view. This is not elective modernism. Elective modernism is compatible with the existence of mysteries that cannot be explained scientifically. It does, however, encourage the scientific study of the observable consequences of the mysteries. For example, one observation that almost every reader of this book can make – and it is an observable consequence of the idea that forms of life are irreducible – is that spell-checkers do not understand language. This is because language is a collective property, not the property of the individuals who make up the collectivity, and no-one knows how to make a spell-checker that can share in collective life. To test the claim, type into your word processor: 'my spell-checker does not understand that I want to misspell wierd in this text'. You'll see that it really does not understand and the problem that cannot be solved is how to get it to understand and keep that understanding up to date in the constant flux of language use in a living society; the only way we humans know how to do it is to take a full part in the society, but we do not know how to make machines that can do that. The observable consequences of the mishandling of language by machines arise from the fact that we do not know how to make machines that fit into human forms of life, and language is a feature of forms of life.[71] As can be seen, elective modernism is a long way from certain kinds of scientism.

Elective modernism is, however, open to the possibility that one day these mysteries will be explained and that 'sub-form-of-life particles' will be discovered. It is simply up to scientists to try if they wish. Indeed, given that open-endedness is a formative aspiration of science, elective modernism encourages the search for more scientific explanations. It is just that elective modernists do not think that an *a priori* commitment to the scientific explicability of all things is a feature of the scientific form of life. That view, on the contrary, seems like scientific fundamentalism. Should there be, however, a person who aspires to explain everything scientifically, their ambition would not be incompatible with elective modernism; their certainty that they or others like them are bound to succeed would be incompatible. The idea that some combination of the theory of evolution, biology and neuroscience already explains all human action is a still stronger form of scientific fundamentalism and is rejected by elective modernism. While in chapter 2 it was said that, under elective modernism, logical positivism lives, it is only the positive side of it that lives; logical positivism's tenet that all non-verifiable propositions are meaningless is rejected.

Elective modernism and the political phase

Elective modernism is concerned with how questions about the observable world are addressed when a matter of public policy includes some scientific or technical issue. Using the terminology of the Third Wave, this is the 'political phase' of technological decision-making. We now set out what the political phase is and what it is not. We begin by showing that elective modernism is quite different from technocracy. Although this might seem obvious from what has already been said, the scholarly context in which we work requires us to make the argument again in order that there can be no room for misunderstanding.

Elective modernism is not technocracy

Technocracy emerged as a social movement in the early twentieth century. Promoted by engineers and scientists, it claimed to provide a rational alternative to the vagaries of market economics. Initially associated with well-known and respected names such as Thorstein Veblen, technocracy's period of mass popularity and formal organization was short-lived, peaking in the 1930s and becoming politically marginal thereafter.

The underlying idea of technocracy has proved much more resilient, however, and continues to play an important role in the discussion of technological decision-making in the public domain. The strength of the idea turns on a pre-1970s, Wave One, model of science in which science produces a superior form of knowledge – facts with a quasi-mathematical or logical level of certainty – that deserves a special status in policy-making. In contrast, non-scientific knowledge, including political preferences and other values, are given less weight because they cannot be justified in a 'rational' way.

Technocracy can be thought to influence technological decision-making in one of two ways.[72] An idealized science and technology replaces politics and technical experts become the decision-makers, planning and organizing societies according to whatever scientific principles the evidence supports. This form of technocracy is rarely found in practice.[73] In contrast, a more moderate form in which experts advise and politicians decide is found in many democratic societies. Also called the 'decisionist model', this form of technocracy institutionalizes a division of labour based on the distinction between facts and values and allows specialist experts to wield significant power.[74] This is because policy-makers work within the constraints set by the experts and _choose from the options those experts provide_. The technocratic element is clear: experts set the agenda and political judgements are parasitic on the judgements of experts.[75]

It is this second form of technocracy that has been the focus of attention within STS. The concern is that, despite what the Second Wave of science studies has revealed about the nature of scientific expertise, democratic systems have become increasingly technocratic in practice through their ever-growing reliance on expert advisors.[76] This idea lies behind well-known arguments that specialist, technical experts are exerting too much power and influence through their behind-the-scenes work of agenda-setting and problem-framing.[77]

Elective modernism accepts this critique while still maintaining a place for expertise. Elective modernism can reject technocracy but value expertise because it sees science as a form of life and not a set of truths produced by the application of a logical and rational process. Focusing on the values that underlie the form of life of science rather than on the content of scientific knowledge is one of the things that makes elective modernism different. Instead of attempting to defend scientific and other technical expertise by reproducing the 'fact–value' distinction, elective modernism distinguishes between scientific values and democratic values – a _value–value_ distinction – and defends science on moral not instrumental grounds. This means elective modernism treats values as ultimately stronger than facts and not as a weaker form of justification. Its central claim is that technological decision-making in the public domain requires institutions that nurture _both_ democratic and scientific values.[78]

Figure 3.1 summarizes the relationship. Scientific values and the institutions that support them are represented by the shaded oval. The work of these institutions will give rise to some policy options and recommendations, which are represented by the dark triangle. The larger clear oval represents democratic values, which partially overlap with scientific values, and which include both formal and informal political actions. Democratic values and the institutions

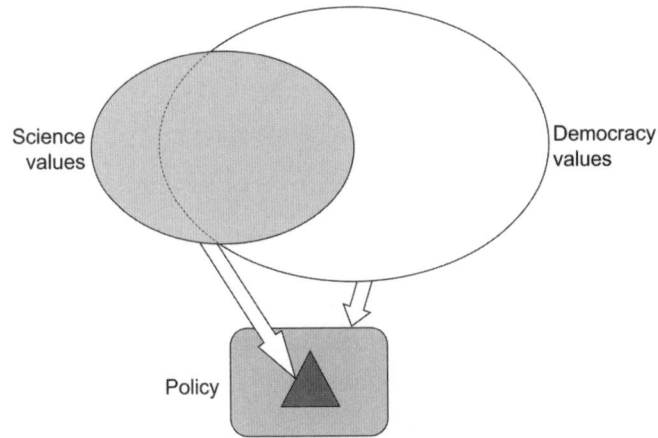

Science
values

Democracy
values

Policy

Figure 3.1 Technological decision-making in the public domain

that support them, along with civil society and public debate more generally, will also give rise to a number of policy options and recommendations – the shaded lozenge – of which those recommended by the experts may form a sub-set. Elective modernism says that, when determining which of the policy options to enact, the policy-making institutions should acknowledge the existence of the triangle clearly and fairly.[79]

What distinguishes elective modernism from technocracy is that recognizing expert advice is not the same as endorsing or accepting it. Under elective modernism, policy-makers are under no obligation to adopt any of the policies put forward by experts, and may reject all of them if they so choose. This outcome is shown diagrammatically in figure 3.2, which shows a legitimate policy outcome in which proposals put forward by technical experts have been rejected. Because political decisions can always over-rule technical recommendations elective modernism cannot be reduced to technocracy.

There is, however, an important rule about how political institutions should over-rule any technical opinion. The rule is that it must be done openly and it must be clearly stated that it is happening. The technical consensus must never be disguised or distorted so as to make the political decision easier. This does not reduce the

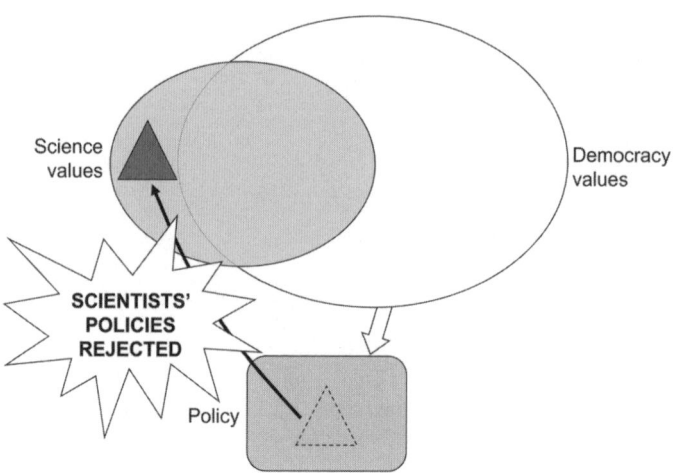

Figure 3.2 Why elective modernism is not technocracy

range of policy options being considered, but it does restrict the kinds of *justifications* policy-makers can offer.[80]

The way this works can be illustrated by Thabo Mbeki's decision not to distribute anti-retroviral drugs to pregnant women with AIDS during the late 1990s because there was a controversy about their safety. In fact, there was no scientific controversy: the concerns Mbeki took seriously were no more than the online presence of a fringe group of scientists whose ideas had long been dismissed in the mainstream scientific community. It is possible that Mbeki had other, more political, reasons for this decision – perhaps he did not want South Africa to fall under the thrall of Western pharmaceutical companies; perhaps he did not think the country could afford the drugs; perhaps he did not want to give credence to the neo-colonialist image of a promiscuous, disease-ridden country. Under elective modernism, these would still be legitimate reasons not to introduce the policy, but these are not the reasons he gave. Assuming, for the sake of argument, that these really were the reasons for the policy, the criticism we would make is that justifying the non-distribution of anti-retrovirals as a result of technical uncertainty about their efficacy disempowers the political process. In this case, it allowed Mbeki to avoid taking responsibility for a political decision by pretending there was scientific controversy, when there was in fact a very strong consensus that the drugs were safe.

There is a clear distinction here between elective modernism and the views of someone like Brian Wynne who considers that Mbeki was (or at least could be) justified in his decision because:

> he was said by others who were closer to his thinking and speaking, to have been saying, not that there is no causal connection between HIV and AIDS, but something very different and orthogonal to the propositional question in itself – that the causal progression of HIV to full-blown AIDS is strongly exacerbated by poverty, malnutrition, immune-system deficit, bad hygiene and sanitation conditions, and other poverty-related conditions, which extravagantly expensive western commercial drug responses (this was before cheaper but still expensive more local generic drugs were available) advanced by extortionate global corporations would not resolve, but would compete with for investment. He was emphasising his view of the need to focus priority on a different set of salient factors in the multi-factorial situation. It was definitely an arguable position; but it was not a superstitious expression of anti-real beliefs.[81]

As noted above, however, even if this is why Mbeki chose to do what he did, this is not how he justified it. At best, Wynne offers a second-hand post hoc rationalization of Mbeki's actions. More importantly, this kind of explanation does not recognize the difference between what may or may not motivate people and the way they speak and act in public life. It is the latter, of course, that forms the societies in which we live, not least because we have no direct access to the internal mental states of others. It is for this reason that elective modernism has no interest in regulating the ways in which people think. Instead, its aim is to raise the standards by which public officials conduct and justify their actions. Elective modernism assumes that experts often disagree on important technical issues.[82] But whether the technical consensus is weak or strong, policy-makers can over-rule it as long as they explain the political grounds that lead them to choose one side or another. That is not technocracy, under any definition.

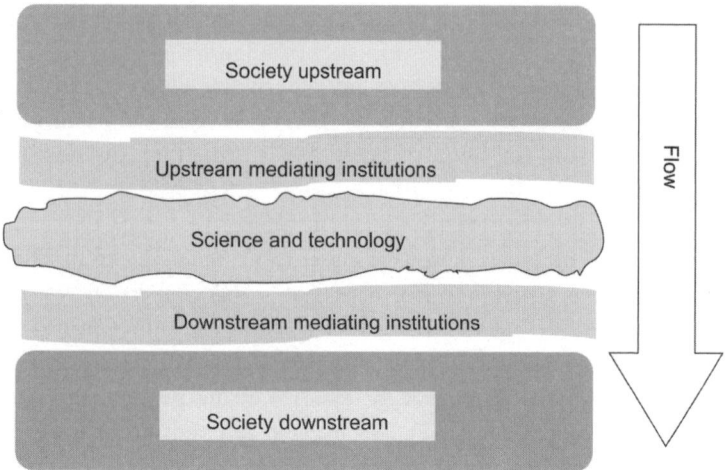

Figure 3.3 The sandwich model of science and society

The sandwich model of science and society

If elective modernism is not technocracy, then what is it? One simple way to see where elective modernism engages with democracy and where it does not is to use a sandwich metaphor (Figure 3.3).[83]

The metaphor separates society into upstream and downstream 'slices of bread'. The filling in the middle is science, shaped and directed by the upstream slice and then feeding its 'findings' and other conclusions into the downstream slice. The sandwich has layers of mayonnaise (or butter) between the bread and the filling. These layers are institutions that mediate between society and science: the 'upstream' layer provides the means by which societies seek to influence, direct and regulate science; the 'downstream' layer provides the means by which the outcome of scientific work is shared with the wider society.

Elective modernism is primarily concerned with the lower parts of the sandwich: the filling and the lower layer of mayonnaise. It does not, in fact, have much to say about the upper layers of the sandwich. It is obvious that societies have to determine what kind of science they want and where scarce resources will be allocated.

Historically, it seems that Western societies and those following their path have chosen to support certain big sciences, such as astronomy, high-energy physics and space science. There is also a marked leaning towards study of the genetic causes of illness, as opposed to environmental causes, and a space-programme-like emphasis on cancer cures. If the authors of this book had their way, they might well make different choices, with the released resources being spent on other aspects of pure science and medicine, but their voices are worth no more than the voices of any other citizen.

Different societies will have different ways of listening to the interest groups or 'publics' that organize around particular social problems and seek to influence how science should be done or how scientific problems should be framed. There is need for the upper mediating institution to sort the expertise-based contributions to upstream decision-making from the local knowledge contribution and the competing interest kind of contribution – the former two being technical, the latter purely political. But elective modernism has little to say about these upper layers, except that much of what other analysts have said about them is not incompatible with it. We might sum up our position as follows:

> there are [always] multiple factors in real risk situations, so then the question becomes which of these are relevant for addressing public interest policy outcomes.... that is not an issue that scientific committees should decide alone. It is a democratic issue, one that should be informed by scientific knowledge but not framed and determined by it. Meanings and concerns should arise within democratic settings articulated through democratic political processes. They should be informed by science, of course, but this is not the same as allowing science to define those public concerns and meanings. There is no reason why something that is democratic and political shouldn't be informed by science.[84]

Note that the top institutional layer must be differently constituted from the lower layer because it will have to include experts on public sentiment and on interest groups and the way they fit into democratic politics. Elective modernism intervenes directly in upstream matters only when it turns out to be a 'club sandwich', which is to say that upstream decisions about how to fund and frame scientific inquiry themselves turn on scientific knowledge. In that case, there is another filling of science and technology above the upper slice and an extra layer of mayonnaise that links its outputs to the other institutions that need this information.

We now turn our attention to the lower layer of mayonnaise and the institutions through which technical advice is made visible and accountable to democratic institutions and their citizens. The guiding principle is that illustrated in figure 3.3 – namely, that the lower slice of bread can deal with the filling as it sees fit, subject only to the constraint that any technical advice received is represented fairly and accurately.

The new understanding of science: the owls

What has happened over the last half-century is that we have learned to reflect on science in a richer and more productive way. Once upon a time, all the rights to the description of science belonged to the scientists, with the philosophers trying to make sense of the mythical world handed down to them. Now, detailed description and reflection on science has become a professional specialism. The clash between how science is described by those with a special training in reflective analysis and those who actually do science from the inside is interesting; it has been mostly misunderstood. Thus, Richard Feynman is said to have scornfully remarked that 'philosophy of science is about as useful to scientists as ornithology is to birds'. For 'philosophy of science', we can read 'science studies in general'. Scornful or not, Feynman was right. Scientists know how to practise science without the aid of analysts of scientific knowledge. The way to go on in science is part of the tacit knowledge of scientists. But Feynman was even more right than he thought: to preserve science as a distinctive form of life, scientists have to ignore, in a determined way, what the reflective analysts of science say. One cannot do good science without disbelieving social constructivism.[85] Individual scientists have to believe they are seeking the truth and that there is a chance of finding it, even while social scientists insist it is the social group that ultimately determines what counts. Furthermore, scientists must ignore the social constructivists if the formative aspirations of science on which this entire thesis turns are to be robust: if it is all social construction, why act with scientific integrity, rather than go straight for some political goal? As we said earlier, we desperately need to preserve the moral imperative that guided science under Wave One, and whatever drives it.

Still more confusingly, then, this need to preserve an old-fashioned view of science is just as important for scientific social science as it is for natural science. To carry out the social analysis of science with energy and integrity requires that the analyst must often ignore his or her own findings in the very act of creating them. When the social scientist is analysing science as socially constructed, he or she has to believe that the truth about social life is being discovered, not that they are putting forward an interpretation for group acceptance or rejection.

The social scientist with scientific integrity has, then, to learn to live in compartments: a compartment for doing science and another compartment for analysing it. This is not so hard because living in compartments is fundamental to training in social science. The social scientist has to learn to 'alternate' between the world of those being studied and the world of the analyst.[86] Natural scientists, in contrast, have responsibility only to their own world. Feynman's claim about philosophy and birds has, therefore, a corollary which he did not discuss: just as one does not ask birds to explain the nature of flight, if one wants to understand the nature of science one does not ask scientists – at least, not scientists, *per scientists*. This is precisely because it is not part of the job of scientists to reflect on the nature of science, but to avoid such reflection.

But, as it happens, not all scientists are the same in respect of their ability to reflect. A few scientists can compartmentalize and analyse their craft as well as, or better than, any social scientist. Wave Two of science studies would not have been the same without the scientists, such as Ludwik Fleck, Thomas Kuhn, Michael Polanyi, Gerald Holton and Peter Medawar, who, as noted earlier, can be thought of as reluctant founders of Wave Two of science studies.

Without Kuhn, Wave Two might never have got off the ground, and without the others it would have been short of certain major ideas or certain rich and crucial case studies. Of course, there are other natural scientists who understand the last half-century of work in social studies of science and have not made major contributions. But even the reflective contributors and the reflective non-contributors together remain a small minority among their colleagues. They would, nevertheless, be a vital resource in re-shaping the institutions that mediate between the world of science and the world of policy. Feynman introduced birds into the debate and we will borrow his metaphor. Natural scientists who also properly understand the social analysis of science we will refer to as 'owls'. Owls are not only wise but can turn their heads almost right round – they can choose to look in two different directions and find it easy to compartmentalize.[87]

There are, of course, many scientists who, lacking any deep reflective understanding, still have something to say about the nature of science in the way 'John Doe' has a view on abstract art. Once upon a time, we thought that working scientists knew all there was to know about science; now we know that the majority of working scientists know very little about it, apart from how to do it. And that, to repeat, is a good thing, because it helps to preserve the values of science. This majority of scientists will be referred to as 'eagles'. Eagles are efficient hunters who find it hard to look in anything other than the forward direction.[88]

Among the eagles are the sharp-clawed scientific fundamentalists; stretching taxonomy, we could call them 'hawks'. Science has achieved much of its esteem through a kind of force: the force of advertising and propaganda. Astronomers' triumphant predictions of the movement of the heavenly bodies gave a huge boost to the reputation of science. When the theory of relativity corrected some of the remaining errors, there was another publicly visible and much acclaimed triumph. Quantum theory has been widely proclaimed as 'the most accurate scientific theory ever', while the cosmological insights of our great thinkers are advertised weekly by scientist-pop-stars. The new quasi-religion, the theory of evolution, is raucously championed, with Darwin as a deity. It is said that evolution can explain not only human form but human behaviour. The hawks even promise to displace the moral leadership of the great religions and fill the gap in moral philosophy with a mechanism. In their way, the champions of artificial intelligence claim that humans are machines and help fill out the details of the new clockwork universe. The hawks rip into scientific heresies with the vigour of religious inquisitions, even employing magicians as rack-masters. Philosopher-apologists – we could call them 'vultures' – digest the carrion left by the hawks, providing a pseudo-academic rationale for the alliance, and too often betraying the very concept of philosophy by ignoring doubts and subtleties.[89]

Advertising science by reference to its performance in narrow domains without complexity is like advertising the national lottery by drawing attention to the winners alone. It gives hostages to fortune and, sooner or later, invites a public reaction because mostly people don't win the lottery, and mostly science cannot solve our problems in the way of Newton, Einstein, Planck and Darwin. Even if the heroic triumphs were truly as they are described in the science story books, they do not bear on the scientific problems that we face in our day-to-day lives. Newton, Einstein, Planck, Darwin and their modern-day equivalents cannot tell us what to do about global warming; they cannot cure cancer, forecast next week's weather, or even predict tomorrow's collapse of the stock exchange.

Furthermore, adopting the iconography of religion – the miracles, the inquisitions, the distortions of reality and the charismatic prophets – momentarily seductive though this may be, subverts the meaning of science even as it champions it. It could even be that overweening scientific fundamentalism was partly responsible for the growth of the 'postmodernist' antithesis.

A characteristic of science's eagles and hawks is that they tend to judge scientific value by results. Just as it is right for scientific mavericks to cleave to their individualistic viewpoints in the face of the majority, it is right for members of the majority to be certain that the mavericks do not have the truth. Unfortunately, this certainty can lead to a sad, backward-reading, analysis of science. Eagles and hawks, unable to accept the varied interpretative flexibility of scientific findings, just know that scientists who find results that differ from their own perceived certainties must have made mistakes.

This is what justifies magicians and the like being brought in to do the dirty work of cleaning up for mainstream science: it is believed to be known in advance that the champions of flawed methods and claims must be charlatans or incompetents and can be disposed of by 'hit men'.[90] This, of course, subverts science as a form of life. Parapsychology, for example, is long past its sell-by date, but the small group of serious parapsychologists who have chosen to spend their lives on what looks like a hopeless task without reward are neither charlatans nor incompetents, they have just made a bad career choice. If they distrust them, mainstream scientists should simply ignore them. It is sad that so many scientists, along with their philosophical outriders, cannot be trusted with these things, but, clearly, they can't.

For completeness, we should mention the bird-like qualities of the social scientists. Those belonging to Wave One are on a continuum with eagles and spend a lot of their time out-eagling them when it comes to attacks on the supposed weaker birds. Those who cleave to Wave Two, and cannot see the point of Wave Three, are also eagles who look only in one direction, though it is the opposite direction from the eagles, hawks and vultures found in the natural sciences, and so they want to replace natural scientists' findings with social scientists' findings. In contrast, those who adhere to both Wave Three *and* Wave Two want to add social analysis to science, rather than replace one with the other. They too are owl-like in that they can turn their heads round and look in two directions. Of course, among owl-like scientists, looking 'forward' and looking 'backward' means the opposite of 'forward' and 'backward' among social scientists but the crucial thing in both cases is the 180 degrees.

Experience-based experts and the underdog

How might this aviary of variously reflective birds combine to make useful decisions? Currently, the most typical kind of contribution made by STS scholars to technological decision-making is representing the public and defending the underdog against the authority of establishment science; this is far less useful than the social science contribution could be. Its value is diminished by the general sentiment within STS – and perhaps the sociological social sciences more generally – for supporting the underdog irrespective of merit, and further weakened by the confusion between experience-based expertise and generalized 'lay expertise'. The malaise is exemplified by social scientists' misguided analysis of the 'MMR vaccine controversy' created by Dr Andrew Wakefield's claim, made during a press conference, that the measles, mumps and rubella (MMR) vaccine was linked to autism.

As a direct result of these remarks, and despite there being no scientific evidence to support the claim, vaccination rates in the UK plummeted over the next few years as parents refused to consent to the administration of the MMR vaccine. The newspapers fanned the flames of public fear by 'balancing' the views of, on the one hand, the mass of epidemiological research showing there was no evidence of a link with, on the other, parents whose children had shown autistic symptoms within a week or so of the injection. Leading social scientists argued that the choice of whether or not to vaccinate a child was not something upon which expert opinion bore more strongly than the feelings of parents and defended those who refused the jab. They argued, *correctly*, that it *might* be the case that, hidden within the statistics of existing epidemiological research, there was a tiny number of children who, for unknown reasons, were vulnerable to the MMR vaccination. But there was no more evidence that this actually was the case than there was evidence for a link between autism and any of the other indefinite number of potential causes of autism that might be selected on a whim. On the other hand, the consequences of vaccination rates falling below that needed to maintain 'herd immunity' were well known: the risk of a harmful measles epidemic would increase and those children who were not vaccinated, and in particular those children with other problems that meant they could not have the MMR vaccine, were at most risk.[91] The social scientists said, nevertheless, that the parents were entitled to demand more research on the potential link with MMR before they surrendered their children to the possibility of vaccine-induced autism. They did not see or care about the disastrous consequences for medical practice and resources if this case set a precedent for the public demanding policies arising out of popular but baseless scares. Bizarrely, these social scientists were demanding more science modelled on Wave One – the parents were entitled to more (perfect) science to settle the matter! The social scientists seemed to have forgotten their cognitive origins in the imperfectability of science and the mythical quality of the Wave One model. Wave Two shows us that science cannot know the world in the kind of detail that would allow us to design a perfect medical and nutritional regime for every individual, and the task is not going to be completed in the foreseeable future outside of science fiction.

We must move beyond this kind of crowd-pleasing deployment of the social science of science: it is far too easy and far too driven by populist sentiment. Fortunately, once it is grasped that science is not something with which democratic societies are constantly embattled but is, on the contrary, a central feature of democracy, it is far easier to escape the damaging sentiments and confusions and make better use of what social science has to offer. Social scientists, often the same ones who were confused, or irresponsible, over MMR, have made valuable contributions in respect of understanding experience-based expertise. There can be more to expertise than is indicated by qualifications and track record. There can be groups who are disconnected from the mainstream but have relevant expertise based on experience, as in the case of the farmworkers concerned about using 2,4,5,T who were discussed earlier; the practical expertise about the conditions in which the chemical was actually used should, according to the story, have been a proper contributor to the debate about how to minimize the risks posed to these workers and their families, though they were ignored by the qualified science community. Whatever the truth of the matter, the story reveals something important about the relationship between the scientific community and the people who actually live with the technological consequences of their work. Users alone know how technological innovations are put into practice so expert users belong to the body who should contribute to the technical consensus from which policy-makers must start the decision-making process. These isolated voices need to be brought to the table, and it is social scientists who are best equipped to do it since science's eagles tend not to hear voices that do not have the backing of scientific institutions.

Science as an oral culture

Another vital contribution emerging from the last decades of social analysis of science pulls in the opposite direction from the widening of the domain of recognized expertise represented by experience-based experts. Social scientists know that science is essentially an oral culture and this means that scientific expertises are narrow. Knowledgeable groups cannot be constructed by using broadcast media such as books, papers and the internet, but only by intense, tacit-knowledge-transmitting, talk, which, by its nature, is restricted to those in face-to-face social contact for a prolonged period. Often such expert groups can be no bigger than half-a-dozen, though in 'big science' they may consist of hundreds.[92] What follows from the narrowness of expertises is that the old-fashioned figure of 'the scientist in the white coat' has no place in the making, or even understanding, of consensus in respect of a specific topic. It is specialists who make and comprehend scientific knowledge, while generalists know no more than lay persons. The internet and the journals – even the mainstream journals – can give a misleading impression of the scientific consensus.[93] Most of what is in even the mainstream journals is ignored by most scientists – the understanding of which items to read and which to ignore can only be acquired through immersion in the oral culture and this, once more, rules out the general public or any kind of generalist.

The notion of science being an oral culture can also stand for the many other new understandings that have emerged from work in Wave Two of science studies. We now know that science is not the set of formal procedures that philosophers tried to describe under Wave One: we know that it is a set of social actions in which the assessment of people and their competence inevitably plays a vital role. This makes a huge difference to the way we understand the meaning of experimental replication, statistical analysis, demands for free access to data and so on. Once more, these things cannot be left to the eagles alone, because the skill of compartmentalizing needed to look in both directions is not something they have ever been trained to develop.[94]

A new institution for policy advice

Social science can, however, contribute far more. Consider MMR-like cases. Elective modernism says that in such cases the public and the politicians should base decision-making on the consensus of technical opinion. But how is consensus to be recognized when most such cases – unlike that of MMR – are typified by deep disagreement among the experts? Imagine some new kind of treatment has been invented by the medical profession. Is it like one of the early and dangerous vaccines? Is it the equivalent of thalidomide or of aspirin? Assume that the double-blind tests have been completed without alarms. We now have all the understanding we are going to get from the scientific community before a decision has to be made. Yet Wave Two has shown us how fallible this understanding is. And, unless the circumstances are very unusual, there will be sources of doubt even within the scientific community.[95] The precautionary principle does not help us because it would simply say 'never use any new drug'. What, then, do we count as consensus?

We have to begin with the technical community within which the controversy is being played out. It is only the core-set scientists plus the relevant experience-based experts with equally specialized knowledge, who can understand the technicalities of the controversy. Whatever policy-making mechanism we are going to come up with will always be parasitical upon the work of these experts in the technical phase. But this does not mean handing over the decision about what constitutes the best science to scientists alone, because scientists are mostly eagles. There may be owls among the eagles, but there will be few of them. It is too big a risk to hand the decision over to mainstream scientists in the hope that the owls among them will be heard: eagles, the huge majority of scientists, have to believe in their work and champion it fiercely; this is not bias, it is the nature of science.

Look at it this way, the decision about what constitutes the best possible current science, for which we can substitute 'the current scientific consensus', should really be made from behind a 'veil of ignorance'. The scientists deciding what constitutes the consensus should do it disinterestedly, and this means that, in an ideal world, the scientists would not know which position on the matter they themselves support. But it is hard to fit veils on eagles; quite properly, eagles are meant to hunt the truth and grip it once they have it in their claws. And even scientist-owls have to spend much of their time acting as eagles. Except in special circumstances, then, the disinterestedness has to come from outside the technical community. Furthermore, it will have wider legitimacy if it is *seen to come* from outside the technical community. The disinterestedness is best supplied by the social scientists, albeit the social scientists who understand science through two lenses, Wave Two and Wave Three, and, like scientist-owls, can look in two directions.

What is being argued for here is a new kind of institution with the job of deciding what counts as current scientific consensus in respect of technical topics in the public domain. This means identifying who is doing good science and/or has relevant experience-based expertise. This is a scientific job and will be carried out in accordance with the formative intentions of science, but it is not a task for natural scientists alone. The science at stake here is the science of the nature of consensus; the science is not the technical topic under discussion, but the substance and strength of consensus about the technical topic under discussion – it is a social scientific fact that is in question, not a natural scientific fact. Under Wave One, or a technocratic model, those who decided on what counts as the best science would have been the scientists working on and in the science in question – there was no social science topic, only a technical topic – but this was a mistake. Deciding who is doing good science on topic 'X' is quite different from deciding what is the truth in respect of topic 'X'. In science, the locus of legitimate interpretation is close to the producers of knowledge, but that means only that the producers are the right people to decide who has done a particular piece of good science and who has done a particular piece of bad science. In the kind of controversy we are talking about, the producers' views on this will be varied – there will be good science on both sides. Some eagles will think scientist

'A' has done the best science and the truth is 'P'. Some eagles will think scientist 'B' has done the best science and the truth is 'not-P'. All scientists will, and should, cling on to their own view, and all will be certain that their view will prove right in the long term. But in the short term, the policy-makers have to decide which scientific view to work with: this decision turns on the social science of science. We need a new approach to expert advice that uses this new understanding to assess both the substance *and strength* of current scientific consensus. The last fifty years of science studies has been about developing a deep reflective and analytical understanding of the social constitution of science – learning to understand how science goes. That understanding should make a contribution to the definition and recognition of scientific consensus, its nature and strength. Under Wave One there was only scientific truth; now there are grades of consensus. Recognizing the substance and strength of current scientific consensus would be the job of the new institution.

Of course, the old question will not go away: 'Who guards the guardians?' Someone has to pick the experts who will make this assessment of scientific consensus. But, like a converging series – guardians, guarding guardians, guarding guardians – the new institution might get us nearer to the solution. This new institution may not be practicable in the kind of society we know with its existing power relations and the almost total dominance of scientific culture by eagles, but what we are engaged in here is a kind of abstract political philosophy: how would things be in a society we were designing from the outset – it is the same kind of question in political philosophy that Rawls tried to answer with his story about prisoners-of-war.[96] Given the licence to invent freely, the members of the new advisory group would be a mixture of owl-like social-scientists and owl-like scientists – taken together we will call them 'The Owls'. The job of The Owls would be to look at the current state of expert knowledge and pass their conclusions on to the politicians, for them to use or over-rule. The Owls, remember, would have the *relatively* easy task of reporting the consensus, not the hard, perhaps impossible, task of knowing the truth. A crucial role for The Owls would be to explain in public why the oral consensus in respect of a scientific dispute is different from the unreliable material that can be found in the literature and on the internet.

Counterfeit scientific controversies

Here is an example of the new kind of contribution that Owls could make; it has already been briefly discussed. Thabo Mbeki, the President of South Africa, decided not to distribute anti-retroviral drugs to pregnant women with AIDS. Mbeki said the following to the second chamber of the South African parliament:

> There . . . exists a large volume of scientific literature alleging that, among other things, the toxicity of this drug [the anti-retroviral AZT] is such that it is in fact a danger to health. . . . To understand this matter better, I would urge the Honourable Members of the National Council to access the huge volume of literature on this matter available on the Internet, so that all of us can approach this issue from the same base of information.[97]

From the study of science as a social enterprise, social scientists, and reflective scientists, know that scientific controversies have a certain pattern. We know that human ingenuity is such that determined critics can always provide reasons to open up doubt about a scientific conclusion, however long it has been thought to be settled. But we also know that, in good sciences, maverick ideas are given a 'run for their money'. Good scientists look seriously at maverick claims, at least for a moment.

In good sciences, maverick ideas that are not completely crazy are tested, occasionally confirmed, but more often refuted, the results being reported in the mainstream literature. After a period of debate, as refutations come to dominate, the mavericks are forgotten. The mavericks – for completely understandable, noble, eagle-like, reasons – rarely accept that they have been refuted. They find, as they can always find, holes in the counter-arguments or counter-experiments. So they try to publish refutations of the refutations. But the mainstream literature becomes closed to them, so they publish in the fringe journals or on the internet. This fringe material is what Thabo Mbeki was looking at. The arguments he was advising his parliamentarians to read had once had a life in the mainstream literature, but the 'run for the money' was over. This is the kind of thing that The Owls could understand and point out in real-time. They can say there is no real controversy going on: it is a 'counterfeit' controversy. To know this does not require an understanding of the science – though such understanding is useful – but an understanding of the social processes of science. To say that something is past its sell-by date is not to say the mavericks are wrong, it is to say that as far as the practical necessity of deciding on the meaning of scientific consensus is concerned, the mavericks are no longer part of the game. From behind a veil of ignorance, even the mavericks themselves would understand the point. The Owls can say that, if politicians want to form policies with the scientific consensus in mind, they should ignore the internet and other maverick sources in respect of 'this' particular scientific disagreement. The statement would be a finding belonging to the scientific social science of science wrought by natural scientists and social scientists working together.[98]

To repeat, it is not the job of The Owls to say which of the arguing parties are right about the science. The social scientists do not have the qualifications: the maverick scientists, even long after they have had a run for their money, are far better scientists than any social-scientist-commentator. In the Mbeki case, one of them was Nobel Laureate Peter Duesberg, who is Professor of Molecular and Cell Biology at Berkeley and winner of many other awards; he is supported by another Nobel Laureate, Kary Mullis. They might turn out to be right. Even the scientist-owls, with their science qualifications, cannot be sure they have fully suppressed their eagle-like biases when they make a scientific assertion on such a matter. But, collectively, The Owls can say, with relative assurance, that, at a certain time, the scientific consensus no longer takes account of what these mavericks have to say. The policy-maker, whose job has to be to make the best decision in the short term – even if in the long term it turns out not to be right – must start with the consensus, not the truth.

The consensus index

Consensus comes in different strengths. Knowing the *degree of consensus* is as vital for policy-making as knowing the substance. In deciding on degree of consensus, there is no escaping the difficult duty of drawing up a scale of 'eccentricity'. We know that the many letters refuting the theory of relativity that every physicist receives, and which are sometimes distinguished by strange typographical conventions, are not to be taken into account as contributing to consensus-making or consensus-reducing within physics. Somewhere, there is a boundary between what is clearly outside the pale and what should be given more consideration. Background research is needed on the structure of scientific disagreement. Research on the classification of scientific mavericks is already underway, with physics as the topic of the research project.[99] Nowadays, findings in physics are often first promulgated on an electronic pre-print server called 'arXiv', to which authors can upload their work before it has been subject to peer review. But arXiv has continuing trouble setting its boundaries: what are serious contributions and what are not? arXiv has had to establish a 'general physics' category for papers which have 'fringe' qualities but still have the other characteristics of a scientific paper and come from faculty members of universities. There are three boundaries within arXiv and at least two more outside it. There is the boundary that defines cutting-edge scientific expertise from run-of-the-mill publications. Then there is the boundary that defines all 'refereeable' papers. Outside of this are the papers reclassified into the low-status 'general physics' category. Still further outside are papers which are rejected, despite their having the form of a scientific publication. In respect of such a paper, a physicist remarked: 'It's professionally done.... The text is pretty good, the equations are mostly explained and the figures are clear. This man knows how to write a scientific paper.' Such papers might finish up in 'fringe' journals or be self-published: of the paper in question, another mainstream physicist remarked '[the journal in which it was published] traditionally contains material that can't get past (or even to) the arXiv stage of publication. It has either been rejected by peer review or expects to be.'[100] The social science contribution is to understand the 'ecology of the fringe', making sure that nothing is too easily discounted, and making sure that wider groups do not think that, just because a publication has the form of a scientific paper and demonstrates some minimal scientific competence, it is something that ought to be taken seriously within the oral culture of science. Outside of this last boundary lies a range of unpublished materials running out to the 'blogosphere' of what are often referred to as 'cranks'. The legitimating role of the social scientists might be most important here – stressing to those who see what looks like science on the internet that it is not necessarily so, despite the carefully crafted appearance.[101]

The research just described cannot lead directly to a decision about what should be 'out' and what should be 'in', but an accurate, social science-based, description of the ecology of the fringe is, at least, a starting point, while the much trumpeted 'peer-review' scarcely touches the surface of the problem. For policy-making purposes, there is no avoiding reaching an endpoint, however rough-and-ready. The Owls would have to grade the strength of consensus from, say, A to E. 'A' would imply complete, or almost complete, consensus, and 'E', little or none. Politicians take a greater risk, and should be seen as having a more burdensome duty of explanation, when over-ruling an A-grade consensus than in over-ruling an E-grade consensus. For example, Thabo Mbeki, though he disguised it, took a huge risk in over-ruling what was an A-grade consensus within the science, whereas politicians take almost no risk in over-ruling an E-grade consensus that emerges from a group of econometricians. Indeed, with an E-grade consensus, the danger is the opposite – that politicians try to justify their political choices by claiming they are based on scientific consensus when there is none, or such consensus as exists is of the lowest grade; exaggerating the degree of scientific consensus is just as politically disempowering for the electorate as pretending there is no consensus when there is. The trouble is that you cannot vote on something that is said to represent a scientific conclusion. The Owls would limit both kinds of abuse of the notion of consensus.

A job description for The Owls

The job of The Owls can be thought of as jumping off from, and widening the scope of, the role of scientific advisors to governments in the UK or elsewhere, and, in particular, the role of high-level scientific advisors such as the UK government's Chief Scientific Advisor. To do their job, which covers many scientific specialties, such advisors must themselves take advice from bodies of scientists. The differences between what is proposed here and the role of current scientific advisors to governments are as follows:

1. Advisors tend to be eagles themselves.
2. Advisors have no special recourse to the social scientific understanding of science.
3. Advisors are unable to grade a consensus in a formal and visible way.
4. Advisors, for these very good reasons, have less public legitimacy than would the new institution.

The Owls would, essentially, do the job of the Chief Scientific Advisor, advising on the substance and degree of consensus about some technical issue. But they would be a statutory committee appointed in a politically neutral way with full scrutiny of the appointing procedure – and their reports and conclusions would need to be a matter of public record. The work of The Owls would differ from existing approaches to expert advice because they would base their conclusions on a wider set of considerations and a wider base of knowledge. When a grade of consensus is set, it will be done on the basis of the science in question brought to the table by the scientist-owls and such specialist eagles as they consult.[102] The grade will also be based on the social science brought to the table by the social-scientist owls and such persons, including experience-based experts, whose contributions they might invite. These social science contributions are, in sum:

1. understanding the difference between policy-making and the determination of truth;

2. understanding and enhancing the visibility of experience-based experts whose role might be overlooked by scientists acting alone;
3. understanding science as an oral culture based on tacit knowledge, and all that follows, including the nature of different kinds of expertise;
4. understanding the way scientific controversies unfold, and the notion of 'sell-by date';
5. understanding and giving more or less weight and more or less legitimacy to the views of those who occupy the borderlands of science.

Inevitably, reaching a conclusion about the substance and grade of scientific consensus will be a matter of debate. All that is being proposed here is that the debate be carried out by the right kind of experts within an institution where science is valued and its assessment has the best chance of achieving political legitimacy short of populism. The work of The Owls will be transparent, and their resulting view of the consensus and its grade will be public.

Under this scheme, there would be no ambiguity about the role of mavericks: their job is to propose, and it is The Owls' and the politicians' job to dispose. For example, it has been argued that scientists who are opposed to the use of genetically modified crops should cease from making their views public, given the potential of the crops to save lives in developing countries. But for a scientist to withdraw endorsement from what is considered to be a true fact about the world just because others think it potentially dangerous is to betray science's heritage.[103] It is both proper, and wiser all round, that scientific mavericks are not tasked with making policy decisions, even when those imply the suppression of their own work.

Under this treatment of the problem, there is also no ambiguity about the relationship between science and politics. One of the most important conclusions that emerged from Wave Two's deep analysis of scientific practice is that science and politics were not only miscible, but always mixed, and, though this remains so under Wave Three, the formative aspiration must be to keep them separate.[104] The determination to treat politics like oil and expertise like water must apply at every institutional level. Oil must be made to float above the water in the bottle of the politics of societies writ large; political decisions cannot be determined by scientific consensus, and nor should the scientific consensus be 'distorted' to fit political preferences. When the bottle is shaken, the oil and water must remain as separated drops in institutions where politicians and scientists work together.[105] When the bottle is shaken harder, even individuals – such as scientific advisors to governments – who start as career scientists but learn how to 'speak' to power, must try to constitute themselves as an 'emulsion' not a solution or a mixture.[106] The duty of the advisor remains to keep the scientific/expert part of the advice separate from the politics of the advice – or, at least, to aspire to do so. Only in this way can the idea of a scientific/expert form of life be retained, and only in this way can scientific advice-giving be compatible with elective modernism. Thus, it must be right for a government researcher to be both a civil servant and a scientist, and to operate according to different normative standards in each of the roles and be judged by those different standards. Whichever level of institution is being discussed, the idea of formative aspirations can be invoked when there is doubt whether to act scientifically or politically.

Above all, politicians must keep things separate. Politicians must clearly and transparently accept any policies that seem to arise from the scientific consensus or, equally clearly and transparently, over-turn them and make their own policies. The Owls won't be setting out to make policy, but where consensus is strong certain policies might seem immanent in the scientific consensus – to overturn these will need political courage, but it will not be The Owls' job to resist. The Owls will be concerned only with making sure that policy-makers know what the current scientific consensus is. To know how risky a thing it will be to reject a consensus, politicians and policy-makers will have to know how strong it is; The Owls will tell them. Politicians, under this scheme, will no longer have the freedom to choose between competing views of experts according to their political preference and still claim they are following scientific advice; they will no longer be able to pretend there is an economic consensus when there is none; they will no longer be able to pretend there is scientific disagreement when there is agreement; and the public will have another body, drawn from a wider base of expertise, to explain the nature of scientific evidence and the processes of science. It will no longer be the job of politicians to pick and choose scientific advice, it will be the job of The Owls, so it will be harder for politicians to choose scientists to fit the preferred politics, while, in compensation, it will be easier for governments to resist ill-informed popular sentiment.

A problem that still needs a solution

All this may be nice in theory – or not so nice, if you believe that Plato's problem of 'Who will guard the guardians?' needs a complete solution before one moves forward – but there is still a *practical* problem about picking the members of The Owls. The difficulty is how specialized, in terms of the science under debate, the Owls have to be. We have talked as though The Owls would be a single standing committee covering all the sciences, but one of the things we know from science studies is that scientific expertises are narrow and deep – there is, to repeat, no such thing as 'the scientist' in the white coat. This creates a real difficulty for recruiting owls, as the membership of The Owls is going to have to vary from topic to topic.

In the case of the scientist members, given the small number of really acute owl-like scientists, we can imagine there would be a few particularly valued members – the equivalent of the Medawars – who would be asked to assume standing committee status with the role of picking the more owl-like scientists, from both – or all – sides of the debate, but working at the cutting edge of the specialty, to complete that side of the quorum. It will be difficult but one can see the solution.

In the case of the social scientists, it will be much harder because there are so few social scientists who have any depth of understanding of scientific specialities; nowadays, the vast majority of social scientists criticize science from outside, not the inside. In the short term, it is likely that this side of The Owls committee will be sparse and not necessarily optimum. For a solution, we need to look to the longer term: we need to turn the social studies of science around, so that it becomes expected that sociologists will treat their case studies in the kind of technical depth that historians take as normal. In the long term, we need more of that kind of social scientist to fulfil that most vital function in society.

We have invented a new institution, The Owls, to make decisions about the nature and strength of consensus in scientific and technical domains. But maybe all that is needed is to adjust a range of existing institutions so that they are populated with Owls drawn from both social and natural science, who understand that their job includes defining the nature of the consensus in the technical areas that they are most suited to survey, and who take as part of that duty the assignment of a grade to the consensus: the grade is as important as the substance of the consensus. Maybe institution-by-institution would be an easier way to introduce these ideas to the society we live in, but it must be much more than a cosmetic change and it might be hard to ensure this in the absence of something radical.

Conclusion

The main work of the book is now complete. We have made our case for valuing science on the basis of its formative aspirations and invented a new institution that could act as a bridge between the worlds of science and policy-making. In doing so, we explained what the technical phase of a technological decision looks like in practice: it is the attempt by scientists and other experts to establish the facts of the matter whilst simultaneously upholding these formative aspirations. The methods used, and the way the work is organized, will vary from case to case, as will the need for experience-based expertise from outside the scientific community. What unites these diverse practices is the common set of scientific values that inform the actions of all concerned.

We have also set out what follows from this idea. We elaborated the idea of the political phase in order to show how science and politics can interact without either being reduced to the other.

Again, values rather than rules or procedures are the key to understanding elective modernism's reach and its limits. Under elective modernism, the political phase is characterized by the supremacy of democratic institutions in a society that recognizes the importance of science as a cultural resource. This means recognizing the importance of the technical phase, but not being dominated by it. It also means accepting that scientists on their own are not the best people to judge the weight that should be given to scientific advice when policy decisions are to be made. The reason is that science is not the truth-making machine of Wave One, but the complex and contested set of social practices described by Wave Two. The new approach to expert advice – The Owls – builds on this new understanding of science to provide policy-makers and citizens with the best available assessment of scientific consensus. What they do with it is, of course, not for us to say; our argument is only that, whatever their choice, it should be informed by the best available evidence. If you agree with this, then you are an elective modernist.

Part III
Academic Context

4

Elective Modernism in Context

The analysis and ideas set out in the preceding chapters draw on a wide range of scholarship which discusses science and democracy. In this chapter, we demonstrate our own commitment to the value of 'continuity' and explain how our work relates to these other debates. The structure and style is more autobiographical than a conventional literature review as we focus primarily on the scholars and work that we have engaged with directly rather than providing an exhaustive and systematic review; this is the academic path we followed.

Elective modernism and the second wave of science studies

Identifying different Waves of science studies simplified a complex and interdisciplinary field into three broad ways of thinking that became salient at different times. Unfortunately, because Wave Three came after Wave Two, many assume that it is intended to repudiate and supersede it. This is not the case. As we wrote in 2002:

An important strand in our argument is to indicate the compatibility of a normative theory of expertise with what has been achieved in Wave Two. The relationship between Wave One and Wave Two is not the same as the relationship between Wave Two and Wave Three. Wave Two replaced Wave One with much richer descriptions of science, based on careful observation and a relativist methodology (or even philosophy). Wave Two showed that Wave One was intellectually bankrupt. Wave Three, however, does not show that Wave Two is intellectually bankrupt. In this strange sea, Wave Two continues to roll on, even as Wave Three builds up. Wave Three is one of the ways in which Wave Two can be applied to a set of problems that Wave Two alone cannot handle in an intellectually coherent way. Wave Three involves finding a special rationale for science and technology even while we accept the findings of Wave Two – that science and technology are much more ordinary than we once thought.[107]

Many have taken Wave Two of science studies to imply that, because science was imperfect and laden with values, imperfection and the endorsement of political values was a legitimate part of science. Raising the status of science and setting it apart from the political sphere is seen as moving back to Wave One. Elective modernism, however, is a development belonging to the Third Wave of Science Studies and there is nothing in the Third Wave that is incompatible with the description of science provided by the Second Wave – the disagreement is solely about implications.

To make the compatibility between Wave Two and Wave Three concrete, we will use the term 'epistemic injustice', which was coined by Miranda Fricker to describe the situation in which a *bona fide* knower is denied that status. According to Fricker, these injustices occur when a claim to knowledge is dismissed because of the social characteristics of the source – for example, evidence may be discounted because it has been produced by women, or manual labourers or Jews – or because the established discourse does not (yet) allow marginalized groups fully to express their experiences.[108] Although the term is not much used in the social studies of science, many of the most celebrated case studies can be seen as instances of epistemic injustice in the sense that their critical edge comes from the realization that experience-based expertise has been excluded or denied.[109] For example, consider the following list of – roughly, Second Wave-inspired – insights and concerns with which we are in total agreement.[110]

In public hearings on environmental matters, peer review or other formal criteria might be used as a standard of reliability for evidence, excluding those with other kinds of expertise.[111] Hearings might be scheduled during the normal working day, making it difficult for those with full-time jobs to attend.[112] An expert committee adjudicating on the safety of a chemical might overlook the fact that the safety equipment and training assumed in their recommendations are not always provided.[113] The number of variables and potential interactions might be too complex to model, leading to simplifying assumptions that seem reasonable to scientists but which do not reflect the world of practice.[114] Findings from research using laboratory animals might not apply to human subjects but be extrapolated nonetheless,[115] or technologies and protocols designed for the average male may be used for females or children who are not suited to them.[116] Forecasts might depend on the predicted values of some parameters whose behaviour is poorly understood, but this uncertainty is not fully considered if the analysis focuses on the central tendency of past data rather than the tails of the distributions.[117] Indeed, it could be said that demonstrating how scientific experts fail to recognize the importance of what they don't know and revealing the tendentious and narrow framing of problems by established experts comprise some of the most important findings to have come out of the STS literature.[118] But all of this is as much part of the Third Wave as the Second.

Equally compatible with elective modernism is the emerging research literature that shows how science is distorted in favour of powerful business interests: elective modernism is, of course, resolute in the face of any distortion of traditional scientific values. We now know that in many policy domains sophisticated public relations campaigns foster public doubts about the status of expert knowledge. Oreskes and Conway's *Merchants of Doubt* traces the ways in which first the tobacco industry, and then the defence and fossil fuel lobbies, recruited retired scientists to critique and undermine scientific research that might otherwise have informed policy-making.[119] In the case of the tobacco industry, the tactics

were primarily those of keeping the 'controversy' alive by exploiting the caveats and uncertainties in research that linked cigarettes
and cancer. Lobbyists and expert witnesses were hired to reveal
'both sides' of the argument to the public long after there was only
one side remaining within the mainstream scientific community.
The same tactics – and, in many cases, the same people – were used
to undermine scientific concerns about the feasibility of President
Ronald Reagan's Strategic Defense Initiative (the 'Star Wars'
missile defence system) as well as a wide range of increasingly consensual scientific research about 'second-hand' smoke, acid rain,
the hole in the ozone layer, and global warming. The mass media
also distort science, though in a less self-conscious way, and this
should give rise to caution when it comes to public opinion in
respect of science and technology.

We might say that the difference between the position being
argued here and the typical Wave Two analysis is that under elective modernism epistemic injustices can be inflicted symmetrically
rather than solely on the poor, the underprivileged and the pow-
erless. Under elective modernism, epistemic injustices can also be
visited by an ill-informed public on elite scientists. Thus, a recent
UK study, which examined public understandings of three major
scientific news stories – MMR vaccinations, climate change and
human cloning – found:

> what people knew usually corresponded with those aspects of the
> science stories that received most persistent coverage. The details or
> subtleties of media coverage are, in this respect, much less important
> than the general themes of that coverage, in which certain ideas are
> repeated and associated with one another. While this does mean some
> information is communicated effectively to most people, it can also
> result in widespread misunderstanding – even if the reporting itself is
> generally accurate.[120]

In the case of MMR, the study found clear evidence of the
distorting effect of the 'balance' within news stories that gave
equal (if not higher) billing to the concerns of parents, despite the
fact that scientific and epidemiological evidence for their claims
was non-existent. Thus, in October 2002, only 23 per cent of

respondents in a representative survey knew that the statement 'the weight of scientific evidence currently shows no link between MMR vaccine and autism' was correct. In contrast, 53 per cent, presumably following the 'balanced coverage' that dominated the media output, thought 'there was equal evidence on both sides'.[121] Similar concerns have also been reported in the Netherlands, this time in relation to HPV (Human Papilloma Virus) vaccinations, where 'a slower-than-expected start to HPV vaccine uptake was attributed to misinformation distributed by groups such as the Dutch Association for Critical Vaccinations opposed to vaccination on the internet and at vaccination centers'.[122] In other words, by exploiting journalistic conventions sceptics can demand that 'both' sides of the story be reported and so create the impression of live and active controversy where, in fact, there is none.

Finally, and somewhat ironically, the sceptics are now developing a new strategy. In the early days of the 'Tobacco Industry Research Committee', the focus was on the science itself, with doubt being created by exploiting the uncertainty and scepticism inherent in scientific work. Nowadays, as Oreskes and Conway show, the debate is increasingly couched in terms of protecting liberty from excessive regulation and state control. Science, particularly the environmental sciences that support the claim that anthropogenic climate change is happening, is now positioned as a threat to the free market economy and hence as a brake on future progress. Science – once seen as the apotheosis of democratic modernity – is now cast as democracy's enemy.[123] One must choose one's preferred society, and elective modernism makes the nature of the choice clear.

Differences from Wave Two

The difference between Wave Two and Wave Three is not about the nature of science. It is about what follows from recognizing that science is a social enterprise and that knowledge is constructed within social groups. The Third Wave argues that different social groups can be distinguished on the basis of the kinds of experience their members have and the norms and values they subscribe to. It also argues that it is possible to differentiate between the technical and political phases of a technological decision, and to identify where the contributions of different social groups and individuals should be directed. In some cases, such as that of 2,4,5,T, this might lead to the conclusion that the technical or political phase, or both, should be more inclusive. In others, such as that of the MMR controversy, the conclusion might be that there is no meaningful technical disagreement and that a political, not technical, resolution is needed.

In contrast, the Second Wave of science studies finds it difficult to argue for anything other than more inclusion. This is partly a result of its intellectual commitment to deconstruction, which is inevitably more concerned, and more comfortable, with breaking down boundaries than with erecting them. It may also be related to the political preferences of those involved. As Jasanoff has written: 'Many STS scholars think that the institutions, practices, and products of science and technology should be characterized in new ways not only for the sake of descriptive adequacy and analytic clarity, but also in order to reorder power relationships: for example, to make the exercise of power more reflexive, more responsible, more inclusive, and more equal.'[124] The difficulty is that this reordering of power often seems to prioritize being 'inclusive' and 'equal' over being 'responsible' or 'reflexive'. It is, therefore, disappointing when Second Wave analysts seem unable to accept the dangers of mass media distortion of science and of the associated technological populism. Thus, the social scientists most closely involved in analysing the MMR controversy refuse to accept that their defence of the public was misguided and dangerous. Instead, they line up with the UK Conservative Party in defending the right of parents to free choice of medical treatments, even though there was never the slightest scientific evidence against the triple vaccine and much to suggest that expensive and highly profitable single-shot vaccines are a less effective public health measure.[125]

Likewise, in the controversy over the disposal of the Brent Spar oil platform, in which a choice needed to be made between disposing of the platform at sea or on land, all parties now agree that the sinking of the platform in the sea was the most ecologically sound method of disposal, while the successful pressure brought to bear in favour of disposal on land resulted in unnecessary harm to the environment. Here, the problem is not so much that technical advice provided at the time, and which was used to support disposal on land, has now been revised. Rather, the problem is that social scientists would still oppose ocean disposal but now do so on semi-religious grounds – it would encourage a kind of 'pollution' resulting in the disposal of more oil rigs at sea and the possible extension of the practice to other kinds of waste, including nuclear waste. It is possible to read this as a debate between the right way to frame the problem and, in particular, whether any kind of utilitarian argument will be countenanced. If utilitarian arguments are seen as relevant, then technical advice matters. If not, and the policy preference is based on a quasi-religious moral stance, then any technical argument about whether or not this particular oil rig will pollute the ocean is irrelevant. In such debates, the concern of elective modernism is simply that framing the matter in solely quasi-religious terms cuts off the possibility of even considering the benefits or harms that might be done to, for example, fish populations by the disposal of one or more oil rigs at sea – while disposal of nuclear waste is a very different matter.[126]

It is not that these approaches to topics such as MMR and the Brent Spar should not be countenanced. Instead, the argument from elective modernism is that the problem should be framed as imaginatively as possible so that all relevant moral positions and technical advice can be obtained. Given this information, the choice of what to do next remains a political judgement that considers this expert judgement alongside other factors, of which public opinion and financial cost may be among the most important. What elective modernism argues is simply that technological decision-making in the public domain should be approached case-by-case and not driven by a pre-determined preference for a particular kind of outcome. In the two cases just discussed, it should have been possible for the social scientists involved – as professional experts on the nature of science – to recognize that the technical consensus was no longer on their side, and to see how this should affect their thinking and that of others in the debate. Of course, this does not preclude them from joining campaign groups and supporting all kinds of political causes in their capacity as private citizens; what we are referring to here is what social scientists can and should say *as social scientists*.

Another, more general, way of posing the same problem, and hence of distinguishing between Wave Two and Wave Three, is in terms of co-production or civic epistemology, both of which are strongly associated with the constructionist approach of Wave Two. Co-production is the more general term and, in the STS literature, denotes the ways in which practices of knowledge-making simultaneously produce both the objects that make up our world and the social institutions and norms that give those objects their meaning. Jasanoff, a leading proponent of the approach, summarizes its insights as follows:

> Public reason is not only an epistemic but also a normative achievement . . . Whenever and wherever reason underwrites and justifies power, it takes its color from culturally grounded understandings of how power ought to be rendered accountable. It is impossible to keep apart judgements of how to know the world in order to govern it from concomitant judgements about how best to govern the world as we know it. Thus, how a democratic society accommodates itself to rule by experts – whether by insisting on individual virtue or by demanding formal technical credentials, for instance, influences the composition of advisory committees, the form and frequency of knowledge controversies, and the means chosen to affect closure.[127]

The application of the ideas can be illustrated using the example of the farmworkers concerned about the safety of 2,4,5,T discussed earlier. In this case, we see the creation of an expert committee (as opposed to, say, a more adversarial, judicial approach) to weigh the arguments and determine the status of different knowledge claims. We also see, in the work of the committee, the co-production of 2,4,5,T as 'safe to use' (an ontological claim about the world) along with the affirmation of social institutions in which scientific expertise is considered superior to that of manual workers (a normative claim that reproduces a social hierarchy). In this case, Wave Two and Wave Three are broadly similar in their interpretation of what actually happened: both agree that the process simultaneously constructs the natural and social worlds of their participants and does so by drawing on and reproducing pre-existing hierarchies of power and status. They also both agree that the outcome was unfair to the farmworkers and that a more reflexive and inclusive definition of expertise that gave more weight to the farmworkers was needed. If such a thing had happened, then both ontological and normative characteristics would have been changed (e.g. 2,4,5,T would be 'dangerous' and manual labourers would have been the epistemic equals of scientists).

In contrast, when we get to the MMR case, the two approaches differ. Under Wave Two, it is entirely possible – and legitimate – to undertake a descriptive analysis that explores how such a large controversy was created on the back of so little scientific evidence. This would say something about (a lack of) trust in science and/or institutions of government, the role of the media, and the power of personal testimony in the civic epistemology of late 1990s Britain. In co-productionist terms, it would describe the further iteration of a cycle in which traditional sources of expertise and authority, including science, are seen as less credible and increasingly risky. Given that Wave Three agrees with Wave Two about how to describe scientific controversies, it would also accept this as description of what was happening. The differences only become clear when one asks what follows from this – when one looks for the 'ought'?

A Wave Two purist would argue that nothing follows: a symmetrical and neutral description has been provided that explains how public standards of reason were enacted in a particular time and place, and that there is nothing more to be said. In fact, what tended to happen was that the Wave Two analysis was used to justify and further stabilize the MMR vaccine as a 'risky' object and to entrench further a normative order in which established institutions are to be challenged and anyone who attracts enough attention in the popular press can be treated as an expert. In contrast, an elective modernist following the approach advocated by Wave Three would argue that, unlike the farmworkers case, the evidence against MMR was non-existent and that the moral and intellectual order created by those in favour of the controversy could not be supported. In short, elective modernism sees part of its role as being to argue for a particular form of co-production and not simply to describe the different forms of co-production that exist within different societies. Here there is a clear divide between elective modernism and much of what Second Wave analysts have taken to be the *consequences of* the revolution in our understanding of the nature of science that has taken place over the last half-century.[128]

As has been explained in the earlier part of the book, elective modernism takes it that the formative aspirations of science should remain unchanged even while the findings of Wave Two are accepted; when doing scientific work, we should cleave to the values of science even though they cannot be realized.[129] The logical imperfections discovered under Wave Two should never have been taken to imply that science was not still a special form of life. Social scientists, above all, should know that forms of life are not collections of quasi-logical rules. Wave Three corrects this mistake by emphasizing that expertise comes from experience and that it is the values embodied in that collective experience which make science an important cultural resource. The philosophy of the three waves and their implications for action within what we have called the technical phase are shown in table 4.1. Note that, under Wave Two, science is eroded as non-scientific values encourage new kinds of behaviour. In contrast, Wave Three retains scientific formative aspirations and so defends the role of science as a distinctive cultural institution that, through the actions of its members, upholds and exemplifies a moral vision. Table 4.1 shows how to act under Wave Three while also showing that Wave Three's philosophy is almost the same as Wave Two's.

Intellectual precursors and contemporaries

Among others, Walter Lippman, John Dewey, Jürgen Habermas, John Rawls and, more recently, Heather Douglas and Philip Kitcher have developed major arguments addressing the problem of how best to balance the role of democratic and expert institutions in the public domain. Here we try to draw out their main

Table 4.1 Three Waves of Science Studies: Philosophies and Actions

WAVE 1		WAVE 2		WAVE 3	
Philosophy	Action	Philosophy	Action	Philosophy	Action
Science is perfect or can be perfected	Nurture science	Science is imperfect and cannot be perfected	Accept and embrace the imperfectability of science	Science is imperfect and cannot be perfected	Act as though science is perfectible. Choose and nurture science
Science is or can be value-free	Nurture science	Science cannot be value-free	Recognize values. Choose and celebrate interests	Science cannot be value-free	Aspire to value-freedom even if it cannot be achieved. Try to recognize interests and discount them
Science is or can be a-political	Nurture science	Science cannot be a-political	Recognize science as a continuation of politics by other means and choose a politics	Science cannot be a-political	Aspire to political neutrality even if it cannot be achieved. Try to recognize political biases and discount them
Findings can be corroborated	Nurture science	Ideals are beyond reach		Ideals flawed	Try to corroborate
Scientific laws are falsifiable	Nurture science	Ideals are beyond reach		Ideals flawed	Try to falsify
Experiments are replicable	Nurture science	Experimenter's regress		Experimenter's regress	Try to replicate
Mertonian norms of science	Nurture science	Norms violated	Forget norms or introduce counter-norms	Norms violated but central	Aspire to norms

ideas, highlighting similarities with and differences from our position.

Political theory and STS

There are now a number of writers who are exploring the connections between political philosophy and the social constructivist view of science put forward by STS. Fuller, for example, advocates a 'republican' model of science in which the expression of dissenting ideas is not simply tolerated but actively encouraged. As noted earlier, however, there is a problem with this approach when it comes to using expert advice in policy-making: in principle, dissent never ends and so, in most cases, a decision will be needed despite the potential for further research. Fuller, as far as we can see, offers no solution to this problem, preferring instead to celebrate the possibility of endless scepticism. As a result, the 'contrast' Fuller draws between the fate of Peter Duesberg (unable to access public research funds since questioning the mainstream view of HIV and AIDS) and of John Bockris (whose research on cold fusion is lavishly supported by private benefactors) is not a contrast at all: both are pursuing science that is past its sell-by date and neither has a strong claim on public funds. Both are, of course, entitled to continue their research if that is what they wish, and it is Professor Bockris's good fortune that the perceived potential financial rewards of cold fusion attract investors.

The other notable attempts to theorize science and democracy have focused instead on political liberalism and the problems posed for liberal democracies by scientific expertise. For example, Stephen Turner, who is perhaps the most well-known advocate of these concerns, argues that 'the twentieth century state . . . resembles an alcoholic who voluntarily gives up his car keys', with expert bodies and commissions providing the guidance and discipline that the individual was unable to muster for himself.[130] Turner sees structural solutions to the problem of expertise as prompting a transformation in the nature of liberal democracy. The original ideal of 'government by discussion' has become almost untenable under current constraints.[131] The result is that political liberalism is faced with two options: a formalized institution of the quasi-judicial kind proposed by James Conant, in which expert arguments are explored and tested in front of citizens or their representatives, or the further delegation of political decisions to expert commissions and, with it, the expansion of technocracy and the decline of democracy.

The aspects of Turner's view with which we agree are that much specialist expertise is inaccessible to lay citizens and that, whilst increasing public understanding about the nature of science is desirable, it won't enable citizens to become experts in scientific fields that affect their lives. The best we can hope for is that citizens become better able to make judgements about what can and should be delegated to experts. We read Turner as being essentially pessimistic on this count, and certainly more pessimistic than we are. His view seems to be that the most likely future is one in which expertise continues to expand its dominion and the ideal of liberalism is directed to ever more marginal issues. No doubt Turner would see our revised advisory committees, The Owls, as yet another expert body that obscures the issues from citizens, and as further confirmation of his view. Were The Owls simply concerned with telling the government what to do, then he would be right, but their role is quite different. The Owls offer a way of making clear to citizens and policy-makers what can be delegated to (or taken from) experts and what remains a topic for genuine political debate, including the possibility of completely rejecting the consensus of experts. They would make clearer – *inter alia* – the nature of expertise and its varying potency.

In this sense we are, perhaps, closer to the work of Mark Brown, who has developed a more nuanced theory of representation that recognizes the inter-mingling of science and politics. Brown argues that the idea of representation conceals a multitude of different aspirations and practices, such that no single forum or process will suffice. Instead, a genuinely democratic system will need to give its citizens 'access to a range of different types of associations, and hence a range of different modes of representation', each of which will need to be evaluated according to its own standards.[132] We do not disagree with his analysis: science and democracy share some of the same resources but have very different goals; one cannot be

reduced to the other and ensuring the integrity of each is important, and so on. Where we differ is that we are prepared to offer
specific policy proposals that, we believe, make both the content
and limitations of expertise visible for publics and policy-makers.

John Dewey and Walter Lippmann

The importance of the Dewey–Lippmann debate of the 1920s
for 21st-century elective modernism is the way in which the two
protagonists theorize the role of the public and its capacity to participate in democratic life.[133] The common starting point is the
fear that democratic systems are cracking under the strain of an
increasingly complex industrial society with a growing mass media.

Lippmann, who is the more pessimistic, identifies two problems: information overload and media inaccuracy. The problem
of information overload arises from the diversity and complexity
of information needed to understand any important social issue.
Given that public debate will always include several different
issues, Lippmann is essentially arguing that no ordinary person
has the time or ability to become expert enough to make sound
technical judgements on each and every one of these issues.[134] To
put it another way, Lippmann argues that technological developments mean the omni-competent citizen envisaged by democratic
theory is no longer to be found. As he wrote in *The Phantom Public*:
'Although public business is my main interest and I give most of
my time to watching it, I cannot find time to do what is expected
of me in the theory of democracy: that is, to know what is going
on and to have an opinion worth expressing on every question
which confronts a self-governing community.'[135]

To make matters worse, the sources that most citizens rely on
to form opinions are inevitably limited and partial. In particular,
Lippmann is concerned about the effects of the inevitable selection and editing of news stories on what citizens are able to know.
Nowadays, the uncontrolled nature of the internet has made
Lipmann's concerns still more pressing as the increasing volume and
diversity of information make the task of filtering, evaluating and
synthesizing the available information ever more challenging.[136]

Of course, neither Lippmann nor Dewey understood the extent to which technical specialists are informed by membership of an oral community and, therefore, they were in no position to understand the systematic differences between 'information-on-the-internet' knowledge and specialist understanding. As Lippmann correctly claims, the ordinary citizen does not have the time to become expert in everything, although by virtue of their work or other experiences he or she may have the opportunity to become an expert in *something*.

From his analysis, Lippmann concludes that the democratic ideal of a fully informed and active citizenry cannot be realized. Instead, he argues for a division of labour in which policy-makers and citizens recognize that only full-time experts are in a position to develop the specialist expertise needed to make properly informed judgements. The role of the politicians and the public is then to make the best possible social judgements about who to trust with this vital work. Given that this delegation risks giving ultimate authority to experts to both frame and solve the problems, it is perhaps unsurprising that Lippmann was seen by his contemporaries (and some more recent scholars) as an opponent of democracy.[137]

John Dewey, in contrast, argues that citizens are capable of a more engaged and informed response to policy-debate. He argues that 'publics' are created by issues; 'a public' emerges as individuals who are impacted by a policy or change organize and respond to this new situation.[138] Defining the public in this way has two implications: first, there are many publics; second, not all citizens need be active members of every public. It also follows that publics may come and go, form and reform, grow and diminish as the issues in which they are engaged wax and wane. From an STS perspective, it is easy to sympathize with Dewey's side of the debate: his 'publics' are in some respects similar to the groups of 'lay experts' mobilizing in response to techno-scientific problems that routinely feature in STS case studies.[139] But there is a danger of mixing up this kind of technical contribution of publics with another kind of technical contribution: specialist local knowledge, including that which can lead to sound local discriminations (see Collins and Evans's 'Periodic Table of Expertises' shown in table

1.1). There is also danger of mixing it up with specialist or local interests, the latter being just the normal to-and-fro of politics and power. The technical expertise that small groups may develop when they confront issues, and the local knowledge that small groups possess by virtue of living or working in a particular place, belong to the sphere of science and technology, but special, vested or local interests, though they may drive the acquisition of technical knowledge, in themselves belong purely to the political sphere.

Still more significant is Dewey's optimistic vision of deliberative democratic citizenship. Although Dewey accepts Lippmann's formulation of the problem, his solution is more direct democratic engagement with the technical experts. For Dewey, it is only through this active engagement that citizens will be able to direct the work of experts and hold them to account. He writes of experts: 'Their expertness is not shown in framing and executing policies, but in discovering and making known the facts upon which the former depend.'[140] In this way, the citizen is enabled and empowered to draw on relevant, specialist expertise to make informed decisions.

The Lippmann–Dewey debate is about how to diagnose the challenges faced by American society. Dewey accepts Lippman's diagnosis of the problem but disagrees about the appropriate response. Lippmann's representative model of democracy involves all citizens, but only sporadically. In this model, elected officials oversee the work of specialist experts who, in turn, provide a necessary counter-balance to the vested interests of the politicians. Should experts fail, then elected officials can dismiss them and, should the elected officials fail, then the voters can dismiss them.[141] In this, elective modernism is close to Lippmann's position. In contrast, Dewey favours a more deliberative model of democracy in which groups of citizens – publics – are always engaged in debates about issues that matter to them. These publics debate issues using information provided by their own and other experts, granting or withholding legitimacy to expert views on the basis of this debate. Seen this way, the Lippmann–Dewey debate is not only a debate about democracy, it is also a debate about how to define the 'public' – 'everyone', according to Lippmann; 'specifically engaged groups', according to Dewey.[142] Both definitions have value but confusion is caused if, for example, specific groups of concerned citizens with a particular stake in the issue are taken to represent the public as a whole.[143] For example, it is too easy to take certain highly visible middle-class groups as somehow being *the* public, whereas really they are only 'a public', often dominated by purely political interests, and this confusion seems to have characterized much contemporary debate in social studies of science.

To summarize, we can say that both Lippmann and Dewey were right. Lippmann was right in the sense that all citizens have a right to contribute, but the link between experience and expertise means that different people must contribute in different ways. At the very least, all citizens, regardless of expertise or experience, can participate in political debate as a matter of democratic right. But Dewey was also correct: in certain circumstances engaged publics can be defined and formed in response to issues that affect them and can offer their views more directly. Whether it is Lippmann or Dewey that is favoured, democratic societies have to find a way to make social judgements about whether to trust certain experts and which ones to trust. It is to inform this contribution to democracy that this book has been written. In so far as elective modernism is concerned with democratic societies as a whole and the place of scientific expertise within them, it is Lippmann who is the more natural bedfellow because of his recognition that the general public can never absorb the ever-expanding volume of specialist information needed to participate in democratic decisions as experts. Where the focus shifts to particular controversies, then Dewey's 'publics' are perhaps more relevant, as long as it is remembered that these groups represent interested elites and other sub-sections of the population, they are not the general public.

Rawls on liberal democracy

The starting point for Rawls's theory of political liberalism is the observation that, in any democratic society, there will be a plurality of views that are held reasonably and in good faith by different groups of citizens. No single one of these views, which Rawls calls 'comprehensive doctrines', can provide the basis for social cohesion as no single one is acceptable to all. This means that there must be an over-arching set of values – freedom, equality and fairness, according to Rawls – which each group endorses on the basis that giving themselves the freedom to act in their preferred way implies that others should be allowed to make the same choice with respect to their own beliefs and practices. The only constraint Rawls's version of political liberalism imposes is that no comprehensive doctrine can violate the core values of freedom, equality and fairness. This distinction between core values and comprehensive doctrines is then reflected in the distinction between the formal political sphere, in which core values are essential, and the diverse background culture made from the tapestry of comprehensive doctrines.[144]

For Rawls, the standard examples of comprehensive doctrines are religions and political associations, but science and secular beliefs are also treated as comprehensive doctrines. Thus, neither science nor secularism can provide a foundation for public policy within a liberal democracy. To make science alone the basis of public policy would, of course, have been a version of technocracy.[145]

There is, however, an ambiguity in Rawls's treatment of science. Rawls, as with elective modernism, distinguishes between political decisions and other kinds of decisions. In his own writing, Rawls is mainly concerned with political decisions that touch on the fundamental issues of freedom, equality and fairness. Elective modernism makes a similar move in defending a formal, somewhat autonomous, political sphere that is separate from the competing interests of the background culture.[146] On the other hand, Rawls does give a special status to science as a form of knowledge within political liberalism, and treats it as more than 'just another' comprehensive doctrine. The different role of science becomes clear in the restrictions Rawls imposes on the ways in which decisions can be made within the formal political sphere. According to Rawls, the only legitimate way for policy-makers and other public authorities to justify their choices is by reference to public reason.[147] This means that 'Citizens engaged in *certain political activities* have a *duty of civility*, to be able to justify their reasons on *fundamental political issues* by reference only to *public values* and *public standards*',[148] where 'public values' refers to the shared political values of freedom, equality and fairness, and 'public standards' refers to principles of reasoning and evidence that all reasonable citizens could endorse. Crucially, these public standards include commonsense, facts generally known, and the well-established, non-controversial conclusions of science. In other words, for Rawls, it is wrong for public policy to be based on, for example, appeals to religious doctrine (at least where these cannot be translated into some form of public reason) but it is not wrong – in fact it is a requirement – that political decision-making be attentive to the conclusions of science.[149] Of course, Rawls did not understand what we now know about science, namely the huge scope for uncertainty and controversy; we might say he was working with a Wave One model of science. We now know that it is difficult for policy-makers to find *anything* that falls into the class of 'non-controversial' science. Under elective modernism, however, the problem is resolved through The Owls, who provide a more nuanced account of the technical consensus that can take the place of 'facts generally known'. This minor modification aside, however, we believe Darrin Durant is right to argue that we have much in common with Rawls's views.[150]

Habermas and deliberative democracy

Like Rawls, Habermas is an advocate of deliberative democracy, but the relationship with elective modernism is a little different. Habermas's account is sociological; for him, democratic decisions can be legitimate only if they can be explained in ways that match the shared understandings of citizens. It must be noted, however, that Habermas distinguishes between 'The informal discussion of issues in an unorganized, "wild," decentered (not centrally coordinated) public sphere that does not make authoritative collective decisions and a more formal political process, including elections and legislative decision-making, as well as the conduct of agencies and courts.'[151] As with Rawls, Habermas distinguishes between citizens' day-to-day political thinking and the process by which collective decisions are made. It is only the latter that is the concern of elective modernism; it has nothing prescriptive to say about what citizens should do or think outside of their participation in formal political roles.[152]

The crucial idea that distinguishes Habermas from Rawls is 'communicative action'. Citizen views become important as they gather support and influence through discussion in the public sphere, creating the shared understandings that provide the foundation of collective political decisions.[153] For Habermas, then, the everyday lifeworld of citizens is at the heart of the democratic system: it constitutes the problems for the formal political process *and* provides the place where solutions are legitimated.

More specifically, Habermas argues that the lifeworld is founded on a network of associations that 'Specialise . . . in discovering issues relevant for all society, contributing possible solutions to problems, interpreting values, producing good reasons and invalidating others'.[154] The complexity of modern society means, however, that responsibility for inventing solutions to the problems must lie with the formal political institutions. They have to respond to the concerns that 'bubble up' from the lifeworld but must draw on specialist expertise to process information and reach decisions, subject to the constraint that they will be seen as reasonable by the citizens. Under Habermas's notion of deliberative democracy, the options developed by decision-makers are a contribution to the deliberations that will lead to a shared perspective from which a new set of priorities and concerns arise. In this way, the definition of needs and what counts as an acceptable solution remain in the hands of the citizens.[155]

That there are resonances between the pragmatistic model of democracy advocated by Habermas and elective modernism's theory of the relationship between expertise and democracy is clear.[156] Consider figures 3.1 and 3.2. As with Rawls, we find a distinction between the formal political system and the wider lifeworld. Habermas's account of the relationship between citizens and the political sphere is richer than Rawls's, but there are no fundamental incompatibilities between any of the models.[157] In fact, both approaches lead to very similar concerns and conclusions.

Habermas's most well-known critique of contemporary society is his concern with the colonization of the lifeworld and the scientization of politics, both of which can be seen as aspects of the technocracy that is rejected by elective modernism. For Habermas, the colonization takes place when the legitimate administrative rationality of the formal political system displaces the communicative rationality of the informal sphere, driving out the organic, informal discourses of citizens, and destroying the citizens' right to create legitimacy for the political and administrative sphere. Though the language is different, elective modernism too places the discourse of citizens over the work of experts, while maintaining the role of the latter. Habermas's colonization and scientization are equivalent to the illegitimate use of a poorly constituted technical phase to address problems that rightfully belong in the political phase. But, as with Habermas, the two different worlds need better and stronger links if the political sphere is to have legitimacy in the eyes of the citizens whose lives it affects.

On the other hand, for elective modernism, the key ideas are that scientific values are recognized as a constitutive part of modern society and that expert advice be accurately represented within policy debate. Beyond being concerned to describe the kind of institution that would represent expert advice best, elective modernism says little about how the process of debate and decision-taking should be orchestrated. To the extent that Habermas is arguing that communicative action is the *only* mechanism that can achieve this outcome, and Rawls is arguing that there are many different versions, then elective modernism is closer to Rawls.

Heather Douglas and the end of the 'value-free' ideal

The idea that science is 'value-free' was once a commonplace of both science and the philosophy of science. Current work in the philosophy of science shows that value-freedom, at least conceived of in an ideal way, is a myth. Criticisms of science policy that might once have been the preserve of science studies are now found with increasing frequency in the philosophical literature. Heather Douglas highlights two problems with the 'value-free' consensus in post-war philosophy of science. The first is that the value-free ideal was never an argument that 'anything goes' in science. Instead, the value-free ideal was about the exclusion of 'non-scientific' values from the selection and evaluation of research evidence. Calling this position 'value-free' is misleading because choices about how to conduct a piece of scientific research are themselves expressions of 'epistemic values'. This leads her to suggest that 'the value-free ideal is more accurately the "internal scientific values only when performing scientific reasoning" ideal'.[158]

The second problem, and the one which Douglas thinks is more serious, is that implementing this modified version of the value-free ideal depends on an assumption that science and scientists are socially isolated. In an argument that will be familiar to anyone within the STS community, Douglas argues that this is both false and unhelpful. In particular, she argues that denying the social context of science discourages critical thinking about the ways in which social values shape science.

In making her argument, Douglas distinguishes between two different ways in which non-epistemic values can inform science: the direct role and the indirect role. In the case of the direct role, values alone provide a legitimate reason for doing or not doing something. The decision of an ethical review committee to prohibit a particular research project is an example of a formal process by which value judgements are applied directly to research practice. More commonly, these direct applications of values are made routinely, and without any self-conscious attention to the ethical position, as researchers unreflectively follow the guidelines of their profession.[159] Elective modernism has no argument with this.

The indirect role is more complex. Here, non-epistemic values work alongside the more traditional epistemic values to determine, for example, how much evidence, and of what sort, is required when the risks of false positives have to be weighed against the consequences of false negatives in order to reach a policy conclusion. In such a case, determining what level of statistical significance or what sample size to use depends on a moral judgement about what is an acceptable risk. Thus, science is saturated by both epistemic and non-epistemic values, a position that is all too familiar to Wave Two social scientists, and therefore one that is, again, compatible with elective modernism. The question is how to act on these insights and this is where the positions diverge.

In exploring the difference between elective modernism and

Douglas's position, a good starting point is to look at the paradigmatic examples of 'policy-related science' that are used to inform the two discussions. Douglas makes it clear that, although her discussion of science is intended to include all science, the relevance of the indirect role for values increases as science becomes embroiled in policy-making:

> Values are needed in the indirect role to assess the sufficiency of evidence. There may be cases of scientific research where only cognitive values are relevant to these assessments, but these cases are merely simplified instances of the norms I have articulated . . . rather than an ideal for which scientists should strive. Such 'pure science' is the special case of science without clear ethical or social implications, one that can rapidly switch to the more general case when research scientists suddenly find themselves facing social implications.[160]

Another way of saying this is that Douglas is dealing with the 'Problem of Legitimacy', which arises when scientists begin to construct policy themselves or do science that is clearly of fairly immediate policy relevance.[161] Douglas, however, is at pains to distinguish between 'pure' sciences, such as gravitational wave physics or cosmology, and the socially relevant sciences of the environment, health and agriculture. In contrast, elective modernism treats all science as the same, along with experience-based expertise. The provenance of elective modernism is the study of controversies in pure science, the way mavericks are treated, boundaries maintained and consensus established within the professional community of scientists. This leads the two approaches in very different directions when it comes to consequences for action, even though the starting points are very similar in terms of the relationship between science and values.

For Douglas, the problem is how to ensure that the indirect role of values is given sufficient weight in the scientific practices that inform policy-making. Her starting point is that scientific advisors must be neither reckless nor negligent when considering the consequences of any error in their analysis. She argues, first, that there is no reason why scientists should be absolved from the general moral responsibilities of care and due diligence that apply to others.[162] Second, having accepted that there is a moral duty of 'reasonable foresight', it is the scientists themselves who are best placed to carry out this task. Thus, it is reasonable to expect a scientist to anticipate what the relevant peer community would agree was reasonable:

> Requiring that scientists consider the consequences of their work does not mean requiring that they have perfect foresight. The unexpected and unforeseen does happen. Holding scientists responsible for unforeseen consequences is unreasonable. What is reasonable is to expect scientists to meet basic standards of consideration, with the reasonable expectations of foresight judged against the scientist's peers in the scientific community.[163]

Despite arguing for the autonomy of the scientific community in this respect, Douglas does not believe that this is enough to ensure that the indirect role of values is given its full weight. The reason is that the moral imagination needed to prompt the necessary reflection is not to be found in the extant scientific community. At least two new institutional practices are needed in order to improve the integration of non-epistemic values into scientific advice. The first is to increase the demographic and cultural diversity of the scientific community itself. The argument here is the standpoint claim that more diverse epistemic communities identify a wider range of tests and that the 'objectivity' they create is thereby increased.[164] Douglas is not persuaded that this is sufficient, however, because to become 'a scientist' means to develop a particular and distinctive set of values. Douglas puts the point as follows: 'Many of the most central values scientists hold are those they developed whilst training to be scientists, and this alone can create divergences between the scientific community and the general public . . . Thus, even a demographically diverse scientific community may still hold values that clash with the broader public's values.'[165]

Thus, even a culturally diverse scientific community could still suffer from a serious problem of legitimacy if its scientific values lead it to frame decisions about evidence and risk in ways that fail to acknowledge what citizens and stakeholders believe to be important concerns. To avoid this problem, Douglas argues for the

increasing use of analytic-deliberative methods through which scientists, stakeholders and citizens can collaborate in order to ensure
that the non-epistemic values that inform scientific work are the
ones endorsed by the society as a whole. Citizens and stakeholders
are able to articulate concerns about what evidence is being gath-
ered, what assumptions are being made, what risks recognized and
so on. This forms the 'deliberative' part of the process. The analytic
part of the process is carried out by the relevant scientific community, which uses its own epistemic standards to assess risks and
take related decisions about evidence and method. Several cycles of
analysis and deliberation may follow, with each new analysis being
subject to further deliberation that might raise fresh questions. As
Douglas explains:

> When judgement is needed in analyses, it arises where there is
> uncertainty that must be weighed. Here the consequences of error
> become important, and must be discussed in a deliberative manner
> . . . The deliberation, if not already begun before the need for judge-
> ment becomes apparent, should involve more than the small expert
> group (as an expert group may have skewed values), bringing into the
> process members of the public. This bolsters democratic accountability
> of expert judgements, but does not threaten the integrity of expert
> work, as the deliberations cannot suppress analyses or require that they
> produce particular answers.[166]

Although the language is different, there is certainly much in
this view that is consistent with elective modernism. For example,
the distinction between the analysis and deliberation seems similar
to elective modernism's distinction between the technical and
political phases. Likewise, the importance attached to the ways in
which technological decision-making frames its technical questions is also shared with elective modernism, where it forms the
basis of the Problem of Legitimacy addressed by much of the
policy-related work in STS.

As noted above, the underlying models of science lead,
nevertheless, to marked divergences in understanding conflict
within science, the role of maverick or fringe science and
the need to delimit membership of the analytic part of any

'analytic–deliberative' process. Like much of Wave Two, Douglas has recognized the importance of values in science but, in doing so, has tended to see more participation as the mechanism for improving their application to technical questions. In doing so, however, she neglects to consider whether or not there is a limit to the influence of this process.

To start with understanding the nature of conflict and disagreement within science, elective modernism has no expectation that the scientists or other experts found in any core-set will be good citizens. On the contrary, core-set disputes are marked by the eccentricities and obsessions of 'larger than life' characters whose certainty that they, and often they alone, have found the truth of the matter is part of the warp and weft of a healthy science. Even without the distant outliers, many or most scientists act as what we call 'eagles', and so they should. Of course, no-one can be against the softening of the boundaries between the scientific profession as a whole and society as a whole – as long as it is a benign society – but one cannot and should not expect every scientist who finds themselves working in some small core-set whose work has social implications to consult citizens and stakeholders about the appropriate balance of risks and benefits before submitting a paper for publication. This is not what scientists do, and nor should they unless they are working primarily as advisors to policy-makers, in which case this kind of work is central to their expertise. That socially significant core-sets should always be populated with socially aware scientists is to expect a lot from the fall of the dice that assigns roles to citizens.[167]

In any case, the calculation of risks and benefits is an enormously difficult and complex job turning on a number of specialist expertises. Worse still, at least from the perspective of policy-makers, the standards found within scientific professions are not necessarily the standards required for assessment of the risks and balances of social change. A good example of the problem is the use of statistics. The statistical standard required for publication in most sciences is the 95% level, which is typically taken to mean that 'there are only 5 chances in a 100 that these results are due to chance'. But understanding what a 95% result means requires understanding of the aggregate activity of entire fields and not just a single experiment.[168] The notorious 'file-drawer problem' arises when many non-significant results pertaining to some effect remain unpublished leaving only the positive 95% results in published literature. The net statistical significance can be zero even though the result of one experiment, or a small set of experiments, is positive. Something similar applies to the 'trials factors', otherwise known as the 'look elsewhere effect', where many statistical cuts are made by experimenters in the home laboratory or other laboratories which cancel out the statistical significance of single or small numbers of positive results.[169] Recently, it has become clear, through the prevalence of failures to replicate what were thought to be well-established effects, that these are real problems not abstract considerations.[170]

The point is that the expertise needed to make assessments of the meaning of an experiment are unlikely to be found within any small core-group of experimenters. This is why elective modernism recommends leaving the assessment of expert consensus not to the existing scientific community but to a new institution that would include this expertise, plus the other expertises that are also needed. Included among these expertises is social science expertise in 'how science goes' that can contribute to understanding which social groups have relevant expertise and how far their practices conform to scientific standards. Take a look at the Thabo Mbeki case – a paradigm case for elective modernism-type analysis. In this case, the crucial problem is how to identify maverick science. Mbeki and his ministers failed to do this because they lacked the necessary expertise, but one can immediately see how an advisory committee such as The Owls, comprising social and medical scientists, could have delivered exactly the 'care, due diligence and reasonable foresight' required to recognize that the alleged controversy was well past its 'sell-by' date. In contrast, a more analytic–deliberative approach would find it difficult to exclude scientists such as Nobel Laureates Duesberg and Mullis. It is hard to see how their views would not have to be taken seriously, and they may well have turned out to be the most persuasive voices in any deliberative institution (as the history of anti-vaccination campaigns suggests). As far as scientists like Duesberg are concerned, they are continuing to act with integrity in their attempts to stop the use of what they see as dangerous interventions.[171] To make a policy out of these heartfelt competing positions requires more than good will, it requires a rationale for finding the 'best decision' in the face of uncertainties that may not be resolved for a very long time, and this rationale will include a high-level understanding of how science goes.[172]

Another, related, divergence with Douglas is that elective modernism associates a far narrower range of values with science *per se*. The danger is that science loses its distinctiveness if its values simply become society's values. This is particularly apparent in the discussion of analytic–deliberative methods, where the over-arching idea is to change the non-epistemic value-judgements of science by aligning them more closely with those of a more democratically constituted body. Under elective modernism, science is championed as a distinct and different form of life that is valuable in its own right. Douglas's position encourages the addition of general social values to those that specifically characterize science, diluting the values that make science what it is.

Were we considering only benign democratic societies, this might be a difference that makes no difference. But, as prefigured in chapter 1, when a society is characterized by what one would think of as abhorrent values, Douglas's analysis has less purchase. Why should we want scientists to take up those values and act as what counts as 'good citizens' in such a society? Historically, we have seen what happens in such cases. Elective modernism too does not isolate scientific values from society, but it encourages society to preserve and cherish science's values – it protects science from society so as to create a division of powers and ideas and even the possibility of a scientific value-led society where society has gone wrong. The tension is exactly that which informed the idea that science could give moral leadership at the time when European fascism was at its apogee.

The elective modernist view has, it has to be said, the consequence that even the most vilely unethical scientific experiments – such as something that involved the torture and death of humans for the sake of scientific knowledge – would still count as science, and science which had integrity, as long as the experiments were carried out according to the values of universalism, corroboration and so forth. Elective modernism would still find such science abhorrent but it would not be because it no longer fell under the definition of science but because it was abhorrent as human activity; it would be science, but immoral science. To give a less extreme but more realistic example, nowadays many people have strong views in favour of or against animal experimentation. Elective modernism has, and can have, no view on the matter. It might be, however, that, should society in general come down against animal experimentation in a clear way, we as individuals living in that future society would feel happier should scientists cease from animal experimentation, but it would have nothing to do with elective modernism. The only change would be in values of the society in which we lived; the newly abhorred animal experiments would not have ceased to be counted as scientific experiments.

To repeat, elective modernism in itself has no ambition to protect every civilized value. As we say in chapter 2, a good society depends on many values in addition to those found in science. Likewise, science cannot, in itself, preserve all social, moral and aesthetic values, and elective modernism does not claim that it should try to do so. Instead, the aim of elective modernism is restricted to safeguarding values such as universalism, disinterestedness, organised scepticism and open-endedness. Elective modernism aims for a science that will provide a cultural resource for the wider society. It is the very insulation of science from social pressures that allows it to be a bulwark for these values, whereas under Douglas's scheme it would become indivisible from society as a whole. In sum, compared to Douglas, elective modernism draws a clearer distinction between the values that define science and those that it borrows from or shares with the society in which it is embedded.

Then there is the question of technocracy. Under elective modernism, technocracy is rejected and democracy trumps technical expertise. Douglas, on the other hand, is more ambiguous. She says that only scientists have the technical resources to provide policy advice but she wants this advice to be supervised by citizens via the analytic–deliberative method. Under elective modernism, in contrast, scientists are also not trusted to make policy, only to make recommendations, but these recommendations are then mediated by a higher body with specialist science-assessing expertise – The Owls. Politics – and hence citizens – enter in the political phase and can reject any technical advice or policy recommendation endorsed by The Owls, subject only to the proviso that this is made clear. The difference is that Douglas would redefine science so as to channel social values explicitly into its day-to-day work as science; this blurs the distinction between the technical and the political phases as extrinsic (i.e. non-scientific) politics can now play an increasingly significant and legitimate role in policy advice offered by scientific experts. Again, the right choice of example brings out the contrast. How would Douglas's model work in the case of economic advice? In this case, it is immediately clear that her approach gives far too much power to economists, whereas elective modernism would make it an easy task to over-rule economic advice for political reasons. This is because Douglas allows social values to influence scientific practice (the indirect values), and has no concept of grading that which emerges as consensus from the scientific community, and does not give the absolute right to the political phase to over-ride even the strongest consensus.

In sum, Douglas's position is close to that of elective modernism at its starting point but diverges as practical consequences are developed. For example, compared to elective modernism, Douglas attaches more importance to the outputs of science, particularly when modified by the analytic–deliberative process. Douglas gives too little importance to the distinctiveness and importance of scientific values as a value system in its own right. She also strikes a different balance between scientists and society, giving scientists (rather than a committee like The Owls) more power to determine the state of scientific knowledge, and politicians relatively less scope to reject expert opinion.

Philip Kitcher

A superficially similar argument for the role of citizen panels also appears in the more recent work of Philip Kitcher. Here, however, the outcome leans towards technocracy. Kitcher's analysis of the relationship between science and society is provided in two recent books: *Science, Truth and Democracy*, published in 2001, and *Science in a Democratic Society*, published in 2012. The overall aim of Kitcher's project is to defend science, which he sees as a rational and epistemically superior form of knowledge, whilst also re-orientating its work towards meeting the needs of society. To this end, *Science, Truth and Democracy* develops the idea of 'well-ordered science', in which citizens are first tutored by experts about what is possible and then choose between these options in order to set priorities. To some extent this could correspond to the role of direct values in Douglas's analysis, or to the role of the political phase in framing and prioritizing research funding under elective modernism.

In *Science in a Democratic Society*, Kitcher is concerned with how such a 'well ordered science' can be integrated into democratic societies. Here he comes close to recapitulating the science court idea, in which experts certify scientific knowledge claims as public knowledge, though this is tempered by a recognition that this process must be treated as legitimate by the public. In a review of Kitcher, Matthew Brown summarizes the twin roles of certification and transparency as follows:

> Certification of a scientific claim as public knowledge requires that the relevant community of inquirers determine that the claim is true enough and significant enough ...Whether a claim is 'true enough' depends on standards having to do with precision and accuracy – how close the results are to the truth, and how likely the procedure is to generate truth. Value-judgments pervade these decisions ... Well-ordered certification requires that these value-judgments pass the test of ideal endorsement. Ideal transparency has to do with whether the public can appreciate the methods of knowledge-production well enough to trust the relevant community of inquirers. Failures of transparency have much to do with the erosion of scientific authority.[173]

Whether Kitcher actually restricts the evaluation of the knowledge claims to scientists alone is a moot point. The phrase 'relevant community of inquirers' *could* be taken to include experience-based experts, but Kitcher does not explicitly say that it does. It is also worth noting that the standard of public certification – that of ideal endorsement – is another Rawlsian thought experiment in which conclusions would be accepted: '*if and only if* they would be endorsed by an ideal conversation: among all humans (and maybe sentient animals), present and future under conditions of perfect mutual engagement aimed at serious equality of opportunity for all people to have a worthwhile life'.[174]

Although it is not entirely clear how this ideal is to be applied in practice, the clues that are given suggest a view largely at odds with elective modernism. In many ways, it seems to have been developed in response to (or at least is most suited to) arguments about climate change and creationism, where, in Kitcher's view, the certification of public knowledge seems to have gone wrong. Thus, for example, although he is broadly in favour of citizen panels, these are framed within a deficit-model way of thinking, in which citizens participate in order to improve their understanding and, we assume, to appreciate why the experts are right. In contrast, elective modernism, as noted above, would see value in citizen engagement in the political phase regardless of whether or not it led to more support for science, and would also endorse the idea of engagement as dialogue rather than simply the top-down information flow implied by Kitcher.

This presumption of expert superiority is even clearer in the sections discussing climate change. Here Kitcher goes beyond the standard technocratic model of experts providing choices and policy-makers choosing between them, to argue that scientific experts should actually just decide. The argument is that, because science is not value-free, these values are best applied by those with the highest levels of specialist expertise, which means the scientists. As he writes: 'Do we really think that *our* judgement – or that of anyone else – would be as good as that of a scientist whose lengthy immersion in these issues leads to the admittedly imprecise assessment offered?'[175]

This position is very different from the minimal default position advocated by elective modernism. According to that view, scientists or experts would never take such a policy decision, and nor would decision-makers be bound to accept even the most consensual expert advice. Instead, decision-making authority would rest with the elected representatives, with the role of expert advisors being to explain the substance and strength of scientific consensus with which the policy-makers would engage in making their choices.

As Mark Brown notes, Kitcher's desire to defend and bolster the role of established expertise leads him to some dubious and contradictory positions.[176] To give one example, Kitcher argues that it would be acceptable for scientists to suppress results that might undermine a policy consensus because these actions could be seen as anticipating (and hence negating) the political mischief that opponents of these policies might cause if they had this information.[177] Such actions are a violation of elective modernism's scientific norms and also offer science's 'eagles' too much freedom to make policy as they see fit; they are technocracy in the making.

Summary

In this section, we have shown that elective modernism is compatible with a number of democratic theories and principles.[178] In making the argument, we have followed Durant in focusing on deliberative approaches to democracy and, like him, we see our approach as being closest to that of Rawls.[179] Like Durant, we also consider elective modernism to be incompatible with approaches that grow out of the politics of identity which treats different 'standpoints' as growing from level epistemological ground. Certainly, different social positions give rise to different experiences and different expertises, some of which are especially relevant in a specific context, but there cannot be different sciences in different social spaces.

Beyond the insistence that expert advice should be provided by experts, elective modernism largely supports contemporary developments in political theory around deliberative systems, in which it is argued that deliberation and representation occur in multiple forms and locations.[180] In this sense, it is broadly consistent with much other scholarship on the relationship between STS and political theory: there is a need to ensure that the classic democratic problems of representation, authority and accountability are addressed not in a single forum but as a process that unfolds over time and space. Elective modernism sees the unfolding relationship in terms of the sandwich model (figure 3.3).

Finally, we have also shown that elective modernism is consistent with much recent work within philosophy of science, particularly that which recognizes the social nature of knowledge and which argues for the value of increased scrutiny of expert knowledge. Where it differs is in the decision to value science on moral grounds rather than epistemic grounds. It also differs markedly from the conclusions for practical action that are taken to follow.

5

Institutional Innovations

The science studies of the last decades has given rise to a surge of interest in new institutions for technological decision-making in the public domain. Heterogeneous forums in which non-expert stakeholders and citizens play a significant part alongside social scientists and natural scientists have attracted a lot of interest.[181] The epistemic rationale for these so-called 'hybrid forums' is that the traditional sources of expert technical knowledge are too narrowly focused to capture all the issues that need to be debated.[182] The impact of this 'policy turn' has been documented in a series of articles by Rowe and Frewer. Writing in 2004/5, they identified over 100 different kinds of engagement mechanisms, which they categorized into a typology of 14 different approaches based on communication, consultation or participation.[183]

It is also worth noting that these calls appear to have been heeded by policy-makers. In the UK, the landmark document is usually seen as the report published by the House of Lords Science and Technology Committee in 2000, though there are several other UK policy documents and proposals that strike a similar, pro-public-engagement tone.[184] More importantly, there is also evidence to suggest that these ideas have been put into practice in the UK and elsewhere. Elective modernism adds a new institution to the menu of innovations. We now set out the menu and compare The Owls with the others on offer.

Citizen panels, juries and consensus conferences

Citizen panels, juries and consensus conferences draw on ideas of deliberative democracy and recruit a relatively small sample of citizens as participants.[185] The recruited citizens learn about the issues from presentations by a series of experts and reach a collective viewpoint by discussing what they have heard amongst themselves. Depending on the procedures, citizen panels may also get the option to recall previous experts, to select new experts and to present reports that contain minority as well as majority viewpoints. The aim is to obtain an informed opinion from a group of lay citizens that can be used within the political phase of technological decision-making as a representation of what citizens with an acquaintance with the issues raised by the science in question would want. Citizen juries are not, therefore, an alternative to the advice of specialist experts, they are an alternative to uninformed public opinion or, as in the case of MMR, populist demands stoked by the systematic misrepresentation of science by the mainstream media.

The best-developed examples of these approaches are often to be found in European countries, particularly the Netherlands and Scandinavia, but there are examples of consensus conferences, citizen juries and similar events being held in Austria, Canada, France, Japan, New Zealand, South Korea, the UK and USA, among others.[186] From the perspective of elective modernism, citizen juries and their ilk would seem to work well as a form of upstream mediating institution. Citizens are enabled to gain some knowledge about the science question, including both its promise and its risks. Clearly, there can be no expectation that these citizens can gain an understanding commensurate with the specialist communities that provide testimony but, as with the legal jury, there is every reason to expect that they can reach a considered and informed judgement that decision-makers can take seriously because of the way in which it has been generated.[187]

Constructive Technology Assessment

Constructive Technology Assessment (CTA) is one example of a more general approach to technology assessment that also includes Interactive Technology Assessment and Participatory Technology Assessment. Although different in some respects, the approaches are united by the promotion of social learning and embedding the concerns of users and citizens into the design process at an early stage. As Johan Schot and Arie Rip put it: 'This family of approaches is characterized by its commitment to what we see as an overall TA philosophy: to reduce the human costs of trial and error learning in society's handling of new technologies, and to do so by anticipating potential impacts and feeding these insights back into decision making, and into actors' strategies.'[188]

The origins of CTA are to be found in the mid-1980s in the Netherlands Organization of Technology Assessment – later renamed the Rathenau Institute – and marked a clear shift in the conduct and use of Technology Assessment. This new approach included both the use of Technology Assessment techniques in decision-making and, crucially for STS, the recognition that these Technology Assessments needed to be embedded within broader social contexts. This, in turn, has the effect of moving Technology Assessment from a 'downstream' activity to an 'upstream' activity, in which users, stakeholders and citizens can exert an influence on the design process by shaping how the criteria used to evaluate the technology are developed and measured.[189]

The approach of CTA and its related forms of Technology Assessment can be operationalized in a wide variety of ways, as the commitment is to a philosophy rather than a method *per se*. Thus, CTA might be accomplished via deliberative workshops, focus groups, town hall meetings, scenario analysis or any other method that allows different groups to express their views. What matters is that end users (and non-users where appropriate) have the opportunity to articulate their concerns when the design is still relatively open and many options remain possible. The idea is that this will allow a more diverse and representative set of values to be reconciled within the subsequent design. In this sense, CTA is not just a new method for doing technology assessment, it is also an argument for a new, non-technocratic politics of technology.[190]

Given the flexibility of the method, it is no surprise that it has been applied to a wide range of topics. It is, however, particularly well suited to environmental concerns and issues around urban design and planning, where there is a clear role for citizens and stakeholders in determining what the design priorities should be. Examples of CTA being used in this way include:

- A wide-ranging public consultation about the development of Rotterdam harbour that was held before the normal planning process. The aim was to increase the legitimacy of any subsequent development by tackling head-on the belief that the government did not take public views seriously. Although not a 'text-book' implementation of CTA, there was a clear attempt to increase the range of participants and consider a wider range of options and concerns, including whether the new harbour was really necessary at all.[191]
- The use of scenario workshops to understand how nascent technologies such as nano-technology will be assessed by potential consumers and users. Here the upstream character of CTA is essential as the technologies in question are still under development and the future pathways are not yet clear. In these cases, CTA provides a method for anticipating what kinds of risks and benefits might emerge and how they might be mitigated. The crucial insight delivered by CTA is, of course, that the risks as seen by citizens or users may differ significantly from those recognized by research scientists.[192]
- The use of CTA methodologies in healthcare to assess the value of new diagnostic tests. Here the dilemma is how soon an innovative technology that shows promising results can be implemented, given the uncertainty that the limited evidence for its efficacy inevitably creates. By using CTA, it is possible to evaluate the introduction of a new procedure at an earlier stage and, equally important, adapt the assessment to take into account new knowledge as it emerges.[193]

From the perspective of elective modernism, there is little to disagree with when it comes to the principles that motivate CTA. Where technology has implications for the ways in which everyday life is conducted, then the expertise needed to contribute to such decisions becomes increasingly ubiquitous. Of course, this is not to say that all citizens are equally capable of performing the statistical calculations needed to model the environmental impacts of different design scenarios, but it is to say that they have a legitimate stake in determining what scenarios should be modelled and how the output of those models is used within decision-making. Likewise, the process of CTA does not require citizens or users to become the specialist experts that develop nano-technologies or genetic tests. Instead, all it requires – and here elective modernists would agree – is that where these new technologies are designed with the intention of being applied in some social context, then it is a good idea to understand how that context works before attempting to change it.

There are, however, differences between CTA and elective modernism when one steps back from the details. First, CTA is about technology not science. Technology is not different from science epistemically, but the relationship between the product and its consumers is different. It is other specialists who are in a position to judge the products of science according to the standards of the discipline, but it is users who decide if technology is doing its job.[194] This is why participatory assessment processes like CTA can work so effectively – the public are expert users of technology, by definition. Other experts, of course, still need to make the artefacts in question. Second, CTA is essentially an upstream mediating institution that deals with the problem of legitimacy by ensuring that a wider range of relevant social groups are able to contribute to the design process.

Citizen science

Citizen science projects can be divided into two types. On the one hand there are those that are initiated by scientists and take the form of science education or outreach projects in which non-scientists are invited to contribute to scientific research. These typically recruit volunteers to collect data that would otherwise be unavailable to scientists themselves, or attempt to harness the power of personal computers to process large volumes of data. Well-known examples of this kind of citizen science include:

- The Cornell Lab of Ornithology, where amateur ornithologists collect data that are then analysed by research scientists working on projects ranging from climate change to avian disease.[195]
- SETI@home, which was the original attempt to use personal computers for collaborative scientific research. Citizens download a program that runs on their PC and analyses radio telescope data in the hope of identifying traces of extra-terrestrial life.[196]

The other kind of citizen science project tends to include elements of protest or challenge. Here the emphasis is not just on science that is 'by citizens', but also on science that is 'for citizens'.[197] In some cases, citizens recruit technical experts to collect the relevant data on their behalf, as in the case of 'science shops'. Others involve established institutions and expertises being challenged to give more recognition to the expertise that citizens possess by virtue of living or working in a particular area. Examples here typically revolve around health and environment issues, and have led to the emergence of 'popular epidemiology' in which residents and/or patient groups collect bespoke, local data in order to provide evidence of harm that official institutions deny.[198] In developing grassroots campaigns, local actors display considerable ingenuity and resourcefulness in finding accessible and affordable ways of generating evidence that will be accepted as valid and reliable by established experts. This, in turn, leads to citizen groups developing new skills and expertise as they learn how to, for example, transform toy robot dogs into mobile sensors that can monitor pollution levels.[199]

Elective modernism would support citizen science initiatives of all kinds. The aim of elective modernism is to promote the values of science as important in their own right, and encouraging citizens to take part in scientific activities is one way of instilling and sharing these values. This does not mean that all citizen scientists are thereby also high-level experts; developing high-level specialist expertises takes too much time and effort. Rather, the comparison is with amateur and professional sports: taking part in sport as an amateur is enough to gain an appreciation of the skill and values that characterize the professional sport, without ever leading one to believe that a world record is within your reach. Amateur science is valuable in just the same way that amateur sport, theatre, art and other activities are valuable: they enrich the lives of citizens and, crucially, sustain important cultural practices; if amateur science can help diffuse the essential values of professional science through society, it must be a good thing.

Public debates and consultations

Public debates and consultations are now a regular part of science-related policy-making and, in the case of the US, are required by law. Public consultations differ from deliberative events as they are more haphazard in their structure and may involve little more than the government issuing a document and collating responses. Alternatively, more structured events may be created, including discussion groups and so on, in order to gain information from particular sectors or demographics.[200] In the UK, the largest of these public debates was the GM Nation? debate, which was held in the summer of 2003 and funded by the UK government on the advice of its Agriculture and Environment Biotechnology Commission (2001). The GM Nation? debate consisted of a series of public meetings, some organized centrally, but with the majority arranged locally. Feedback forms were collected at these meetings and there was also a website where individuals could complete and return the same form. In addition, a 'Narrow but Deep' set of focus groups was convened to run in parallel with the self-organized discussion groups, and an independent evaluation was funded by the Leverhulme Trust. In general terms, the participants in the GM Nation? debate were fairly sceptical about the benefits of

GM crops, though other samples (e.g. the Narrow but Deep focus groups and the large-scale survey used in the evaluation) reported more ambivalent findings.[201]

Research in STS has demonstrated that many of these forums are deeply problematic. As noted in chapter 3, the way the question is framed, what is permitted to be discussed, how views are solicited, and so on, all influence how the public is able to have its say.[202] There is also the suspicion – reasonable in many cases – that the purpose of these exercises is not so much for those in power to listen to the concerns of citizens and respond to them, but more to gather the information needed to improve the public relations campaigns that will be used to legitimate an already agreed policy.[203] Elective modernism shares these concerns but also emphasizes the potential limitations on the ability of lay citizens to contribute meaningfully to the technical aspects of such debate. There is a real risk that events like these do little more than give the public a chance to air their prejudices. For this reason, we argue that, if informed opinion is required, then, at best, more deliberative forums are needed, but whether these will resolve the problems is still not clear.[204]

Public engagement with science and technology

Although not, strictly speaking, part of a technological decision-making process, the nurturing of generalized public engagement with science and technology is worth mentioning. In Scotland, a long-term medical project known as Generation Scotland, which aims to create a genetic database with a particular emphasis on data that link family members, includes researchers from both medical and social sciences and describes itself as a 'partnership with the people of Scotland'.[205] Of particular interest in this context is the role played by social scientists in the project and the emphasis placed on public consultation, both in the run-up to the formal launch of the project and as it continues.[206] The utilisation of publicly generated data has long been part of science and elective modernism per se has nothing to say about it.

Experts as policy advisors

In most democratic systems, experts are used to advise policy-makers. Though decisions are sometimes delegated to bodies of experts, mostly they only advise and it is governments that decide.[207] As critics have argued, much of the traditional thinking about these issues has been informed by a Wave One model of science.[208] By this, we mean that the involvement of experts in the policy process is typically seen as unproblematic, with controversies and problems regarded as failures of public understanding, rather than of institutional design. Policy responses then take the form of attempting to reduce the public appearance of controversy so that science's natural epistemological authority can be re-asserted.

A clear example is the 'science court' mentioned in the previous section. As originally proposed by Kantrowitz, the aim of the science court was to provide politicians with consensual scientific knowledge, even though the policy-relevant technical knowledge was still controversial within the scientific community. Conceived as a quasi-judicial institution, active scientists were to act as advocates, while mature scientists with diverse scientific backgrounds would take on the role of judges.[209] The consensual advice generated by the science court then fed into the political decision. The crucial difference between this and elective modernism is that Kantrowitz assumed that the outcome of this process would form the basis of any subsequent policy: 'It has occasionally been maintained that scientific and non-scientific components of a mixed decision are generally inseparable. It is, of course, true that a final political decision cannot be separated from scientific information on which it must be based.'[210] In contrast, under elective modernism there is no requirement that policy-makers should follow the expert consensus when reaching their decisions – only that they confront the matter squarely, honestly and visibly.

A more recent argument that includes a similar concern with advising policy-makers about how best to use science is the 'honest broker' model proposed by Pielke Jr.[211] This is one of four science ideal-type advisor roles distinguished by way in which experts respond to policy-makers. Drawing on similar assumptions to Kantrowitz, the honest broker is typically an interdisciplinary group that provide policy-makers with a range of options in the expectation that they will choose one of them. Whilst sympathetic to Pielke's concerns, elective modernism is different in some important ways. For example, under elective modernism there is no requirement that experts provide policy recommendations – their primary responsibility to summarise the expert consensus and so have some affinity with Pielke's 'science arbiter' – and policy makers can reject expert advice if they so wish. On the other hand, one important consequence of the Owls, and one which Pielke would support, is that they make it harder for what he calls 'stealth issue advocates', that is experts who miss-represent the state of expert knowledge in order to advocate for specific policy options, to succeed.

Conclusion

It could be that some of these forums do little more than create the impression of participation and act to head off genuinely radical alternatives.[212] Nevertheless, more participatory and deliberative institutions are a good thing: the early STS critiques of public inquiries and their 'capture' by vested interests remains relevant.[213] Elective modernism is distinctive in its clear separation of science and politics. In the typical STS-inspired discussion, the hybrid forum idea is used to muddy the boundary between the public and experts. In contrast, elective modernism divides their distinctive roles and duties in a clear way. The Owls would involve experience-based unqualified experts who already contribute much to the current set of hybrid institutions, but would not include the general public. The public would still participate in upstream deliberative forums, no doubt similar to the ones described above, but now assisted by an informed assessment of the expert consensus and its limits, and, of course, they would continue to play their role in the normal downstream political processes that would, in the final instance, determine policy.

Part IV

Manifesto

6
Elective Modernism and Democracy

We have argued that democratic societies should start with scientific knowledge when they make decisions which turn on science. They should not finish decision-making with scientific knowledge because democracy always trumps science and technology but, should they choose to over-rule any obvious policy implications emerging from science and technology, democratic societies should do it clearly and accountably; democratic societies should never ignore science or distort its claims for the sake of making a political decision more readily acceptable. The political sphere must dominate science and technology, but that domination implies the unflinching acceptance of political responsibility and it implies accountability to the electorate. Accountability to the electorate must never be diluted by misrepresenting the findings of science.

But there is much more to the relationship between science and democracy than this. Sciences' values are democratic values. Science is important to democratic societies because it supports democracy through its very existence. It is, and certainly should be, a professional institution that leads democratic societies by exemplifying their way of being. Just by being science and acting out its form of life, science gives substance to the way of being of democracy. Of course, there is much that comes under the general heading of 'science' that does not fit the ideal model, but we believe there is enough to maintain hope, and there are analytic reasons based in the ethos of science – which is to make truth, not money or material goods – that encourages one that it has an almost unique potential to fulfil the ideal.

A society based on science is not a new idea, but there are as many dystopias in the literature as utopias. Here the substance of the exercise is close to the implicit democratic politics of Merton's norms of science, though we start with the norms rather than the science. It is also close in spirit to Popper's analysis which uses a falsificationist model of science to argue against theories of history that lead to the supposedly scientific redesign of societies.[214] There is nothing here of John Desmond Bernal's utopian science-based socialism, nor of Aldous Huxley's, dystopian, *Brave New World*. It is not being claimed that science can either provide us with the cornucopia of goods and understandings that will fulfil our material needs and solve our mental and physical ills, or provide a way to design a society from the ground up that will assuredly turn it into a utopia, even if a few eggs are broken in making the omelette. We have woken from these dreams into a less exciting but less dangerous world. There is also no fear of science destroying our humanity.

A good society would, of course, depend on many kinds of values other than those associated with science; there is a wider range of values than those exhibited by science and there are the aesthetics of taste and manners. While there is overlap between the moral values of science and of democracy, a good society needs also to draw from aspects of religion and a range of those secular institutions whose values have not yet been eroded. In this book, we concentrate on the common ground between the values of science and democratic values.

As we have explained, the godfather of the idea that there was an overlap between science and democracy was Robert Merton. Lines 8, 9 and 10 of table 2.2 are the Mertonian norms which make a natural match with the values of democracy. Readers of this book, relaxed in the acceptance of the formative aspirations of science, should be horrified when they hear that a scientist

suppresses another's work because of their race or creed (universalism), or suppresses another's work because of their own special
interests (disinterestedness), or refuses to expose their work to
criticism (organized scepticism) or refuses to make it public (communism). How could one prefer one's *science* to be otherwise, and
how could one prefer one's society to be otherwise? A society
characterized by furtive secrecy is likely to be authoritarian, with
control of information and knowledge being tools of power. Only
dictatorships or the like seek to control all information and outlaw
criticism of the rulers. A democratic society has to be open to
removal of the government if the people vote against it or criticize
it enough to force a vote. George Orwell's *1984* is the fictional
example of a society that turns on secrecy, but North Korea provides a real-life exemplar. In contrast, democratic societies must
favour accountability and transparency within the public sphere
– both of which depend on some degree of communism. Equally,
then, how could one prefer one's *society* to be other than universalistic? Can one imagine a democratic society that divides people
along the lines of ascribed characteristics? There are plenty of societies that do this but the readers of this book, at least the elective
modernists among them, would not describe them as either good
or democratic societies. The case is not quite so clear for disinterestedness. It is essential in science and that is why we are rightly
horrified when we learn of scientists being purchased to produce
findings tailored to the interests of tobacco or oil companies.[215] On
the other hand, it is possible to imagine that the job of a democracy is to balance special interests. Still, it is surely not the job of a
democracy to run the country for the sake of the special interests of
the members of the government; this is called 'corruption'.

Continuing with other values, the design of democratic societies is never ending; they are never finally fixed according to some
plan (line 16 of table 2.2). Democracies can always be improved
and they must always be allowed to change, so democracy is open-
ended. For open-endedness to mean something, democracies must
also be individualistic in the sense that individuals must be allowed
to stand out against the mass (line 14). In a good society, one must
be prepared to listen to anyone, irrespective of race, creed or social
eccentricity. One must accept that a heretical voice might be right

and the majority wrong, though how one should act on this is not so clear. A good society cannot exist without Kuhn's 'essential tension' – the tension between the individual and the majority – in its politics as well as its science. The society may be governed by consensus, but consensus must always be open to challenge and change by individuals.

Finally, there must be a role in democracy for expertise or why would anyone invest in improving their skills (line 2)? Because good analysis in any sphere demands high levels of craft-skills or experience, some are more capable than others at both producing knowledge and criticizing it. This means that a democratic society must preserve a special role for experts, and for experience more generally.[216] A good society will make the views of bodies of experts highly visible, even though it may over-ride them politically. That said, experts are experts only in their narrow domains, and this realization is a further bulwark against technocracy: there are no specialist experts in the making of political choices.

A practical consequence for democracy follows because the scientific and technological expertise which, under elective modernism, is a vital component in technological decision-making can make its contribution only under certain kinds of social organization. Therefore, elective modernism favours a certain *style* of democracy: democracy must nurture the generation of the evidence, the consideration of evidence, and political accountability for the use or misuse of evidence. There must be time for evidence to be generated, time for the meaning of any current scientific consensus to be weighed and acted on or rejected, and a period for the outcome to be judged before politicians are made to account for their decisions. There are many versions of democracy but elective modernism favours systems that allow the time and space for technical debate, and time and space for the output of technical debate to be exposed and considered. Thus, elective modernism and the Third Wave are not only against technological populism but against too much populism of all kinds. Science and technology take a very long time to find the truth but they also take time, albeit much less time, to form the insecure consensuses which feed into political decisions. This means the institutions must isolate technical consensus formation from immediate political preference, something that becomes terrifyingly obvious when societies are not benign.[217]

So in these six matters – universalism, disinterestedness, organized scepticism, open-endedness, the tension between individualism and popular opinion, and the valuation and nurturing of expertise and experience – there seems to be an overlap between a good, democratic, society and the formative aspirations of science. Elective modernism can, then, be described as a kind of 'scientism' – a scientism that holds that science should be treated not just as a resource, but as a central element of our culture.[218] But elective modernism is far too deeply informed by the last half-century's critical analysis of science to be anything other than resolutely opposed to any other kind of scientism.

What this book has to achieve is the establishment of the central role of science in democracy. It has tried to undo, to some extent, the erosion of science's cultural status that has been a constant motif since the 1960s.[219] But we have also tried to answer the practical questions that follow on any re-arrangement of the cultural terrain and not simply go back to the 1950s and earlier. Who decides who is a scientific expert? What do we do when scientific experts disagree? How do we design decision-making institutions that take science into account in a sound way, without creating technocracy? Our solution to this is The Owls. Disappointingly, it is just one more committee but, more interestingly, it is a committee that melds the social scientific understanding that has been generated over the last half-century together with science, so as to solve the social scientific problem that needs to be solved for science and technology-related policy-making: what is the current consensus within the scientific community? This isn't a scientific problem because it isn't asking, 'What is the scientific or technological truth of the matter?' It is asking, 'What do scientists and technologists currently believe and how firmly do they believe it?' After this question has been clearly and publicly answered, the rest is politics. As contributors to the work of The Owls, social scientists will do more than uncover the problem, they will be part of the solution.[220]

The foundation of the entire argument is as thin as air – a preference for democracy and a preference for the values of science with an acknowledgement that there is a strong overlap between the two sets of values. The foundation is a choice without further justification. If this choice is not compelling, then the book will not be compelling. Furthermore, in so far as the choice rests on a certain description of the form of life of science that is at variance with the way much of science is practised, the entire argument is vulnerable to the charge of naïveté. Worse, it is a naïveté that grows out of turning the gaze away from the many brilliant detailed studies of science that have characterized the last half-century of 'science studies'. What we are grasping for is the possibility that, unlike so many other professional institutions, science can escape from the erosion of its values in the face of the ubiquitous financial and political pressures. It is a naïveté based on a conception of method that separates the day-to-day activities of scientists into two types: those that are 'accidental', and those that are formative of the way of being in the world – the 'formative aspirations'. It is a naïveté based on an argument about two characteristics of these formative aspirations. First, that science's whole purpose is the search for the truth, which means that, though more and more cases of fraud are uncovered and though there is more and more distortion of science's activities brought about by the lure of mundane reward, we can say of science, in a way that we cannot say of most other professions, that to allow this kind of distortion is to cease to be doing science. Once we could say this of art and religion too, but it is not so certain that we still can. The search for truth is, however, integral with the very notion of science so those for whom science is, and remains, a vocation are bound to see any other goal as a negation of their existence. Second, whatever we social scientists say, scientists are sure they can, eventually, find the truth of the matter, if their search is long enough and assiduous enough. As long as scientists believe this, then their methodology demands that they preserve their value system.

Why bet on naïveté?

Is all this nothing more than a pipedream in today's world? This passage is being drafted eleven days after the triumphal announcement of the first detection of gravitational waves. Collins has followed the attempt to detect gravitational waves almost from its beginning in the 1960s; he began fieldwork in 1972, and has continued ever since. That science has exhibited the characteristics belonging to the naïve description almost without departure for the whole forty-three years he has been watching it. The number of less than honest acts he can think of – a bit of over-hyping here and there, and a bit of deceiving journalists anxious for a premature release of the story in the last few months – is tiny. All the rest, even those empirical claims that now look bizarrely optimistic, were based on a driven search for the truth. Why else would one make them?

The contrast between a community ready to proclaim that they are doing science and ready to base their judgements on what they count as a scientific act, and a community that is actively trying to dissolve the boundary between science and politics, is one of the things that drives this book. Among social scientists who study science, one will hardly ever encounter the justification, 'this argument is better because it is more scientific', and is far more likely to encounter acts of knowledge-making that are justified by their (sometimes implicit) political stance. Academics are always fighting to retain the boundary between science debate and 'science war'. Science debate is directed at finding the truth; science war is directed at winning the argument in a lawyer-like or politician-like way. Science debate must start with as complete an understanding of one's opponent's position as possible, and finish with an attempt to show, to oneself, *and reveal to one's opponent*, where he or she is wrong (however hopeless a task it is). Science war is directed not at the arguments of the opponent, but to an outside audience, and misrepresentations of the opponent's position are a useful tactic, whereas it is pointless in science debate. In the social studies of science, it is especially hard to maintain the distinction *because* science is seen as persuasion not truth-seeking.

Science may have been 'deconstructed' since the 1960s, but the argument of this book is that we reject scientific values at our peril. And since rejecting scientific values is perilous, we ought to start proclaiming them once more. A value observed in secret is more fragile than a value proclaimed and celebrated in public.

In most ways, this book is simply a reiteration of what we have always known about science but have lost the will to say. The new thing is the way of saying it. The proclamation is directed to social scientists, but also to natural scientists. Though we have found, literally, that there are bodies of natural scientists who still exhibit and proclaim scientific values, we know it is far from universal in the scientific community. What we are asking is for scientists, as a profession, to take a leading role in society and that means recognizing their vocation for what it should be and rejecting the call from more mundane institutions; it is a heavy burden to bear.

This book is about science and scientific values, but its overall argument has universal implications. These are that the implacable logic of relativism does not lead to deep practical conclusions. The right and the wrong are stronger than logic, and a relativistic approach to analysis should not be taken to imply that every way of living is as good as every other. Above all, this book comes out of the recognition that, in the last resort, there is nothing that can save us from making choices. As academics we spend our lives looking for justifications and reasons for doing and thinking 'this' rather than 'that', and in the small matters we succeed. But academic conference after academic conference, war after war, and atrocity after atrocity demonstrate that, where the really important things are concerned, reasons will not convince and choices have to be made. This book is written to expose the nature of one of those choices and to provide a defence of societies that place the scientific way of life at their core, for reasons that are deeper than science's utility. As can be seen, the key to knowing that one is thinking this way is to look for a defence of science that works even when its outcomes are flawed and useless; the trick is to find a defence of science that works *even when the science itself does not*, and to find a role for science that bolsters democracy without usurping it.

Notes

1 There is some resonance here with Thomas Kuhn's notion of a paradigm which, when defined as a 'disciplinary matrix' (Kuhn, 1970), includes values. The specific examples given by Kuhn relate to what we might call 'internal values', such as accuracy, simplicity and consistency. Kuhn argues that these values are 'widely shared among different [scientific] communities . . . and do much to provide a sense of community to natural scientists as a whole' (p. 184).

2 See, for example, Durkheim's *Professional Ethics and Civic Morals* (1958), in which the professions play a crucial role as intermediate institutions that link the individual and the state. Parsons (1991) also stresses the role of professions, arguing that they demonstrate the possibility (and importance) of social action that is not driven by individualistic or economic values.

3 The quote is from Hanlon (1999:121); he is writing about changes in the legal profession but the point generalizes to other settings in which 'professionalism' is used as a strategy to promote and/or legitimate organizational change.

4 In doing so, we provide a constructive response to Helene Sorgner's assessment of our previous arguments as 'utterly useless for improving democratic institutions' (Sorgner 2016:5), though we should also

point out that these remarks are made in the context of generally favourable assessment of our work.

5 The remarks were made in a speech to the National Association of Science Writers in 1954.

6 For example, Thomas Kuhn never supported the ways in which his work was taken up by what we call the second wave of science studies. The same is true of Peter Winch, whose interpretation of Wittgenstein's later philosophy provides the inspiration and foundation for almost all our own work.

7 These include Collins (1975, 1985, 2004a, 2013a, 2017).

8 The positive reply is Gorman (2002), the three negative ones are Jasanoff (2003), Rip (2003) and Wynne (2003). Our response to these replies can be found in Collins and Evans (2003).

9 For example, the idea of interactional expertise seems to be widely accepted.

10 The extended classification of expertise appears in our book *Rethinking Expertise* (Collins and Evans, 2007). The idea of interactional expertise is developed in a series of papers: Collins (2004b, 2011, 2013b); Collins and Evans (2015a). The Imitation Game method, which is used to investigate the nature, content and distribution of interactional expertise, is explained in: Collins et al. (2006, 2015); Evans and Collins (2010); Evans and Crocker (2013); Collins and Evans (2014); Wehrens (2014).

11 *Democracy and Expertise* is Fischer (2009). The exchange of papers in *Critical Policy Studies* can be summarized as follows: the initial paper is Collins, Weinel and Evans (2010); the replies are Epstein (2011), Fischer (2011), Forsyth (2011), Jennings (2011) and Owens (2011); our response to these replies is Collins, Weinel and Evans (2011).

12 For more details of this example, see Irwin (1995).

13 Both the problem of legitimacy and the problem of extension were introduced in Collins and Evans (2002).

14 We interpret democratic process quite broadly so that street protests and demonstrations would also count as a means of democratic participation.

15 The alternative, which would be used in a Wave Two approach, is to see expertise as a relational or network property in which the awarding of expert status is the thing to be explained. An overview

of this approach can be found in Carr (2010), with Eyal (2013) providing a recent example of the genre.

16 A more recent review of the idea of contributory expertise can be found in Collins and Evans (2015a).

17 Evans (2011) examines the kinds of expertise needed for different kinds of decision-making institution to function; Evans and Plows (2007) looks more specifically at the role of lay citizens (i.e., citizens with no specialist expertise) in deliberative forums.

18 This is first developed in Weinel (2008), but is also discussed in Collins, Weinel and Evans (2010).

19 Collins (2014a) provides an accessible introduction to these ideas. Collins, Evans and Weinel (2016) provides more formal treatment.

20 We are, of course, aware that the term 'modernism' can have a much more complex meaning than this and can be used to signify much more than a simple association with science.

21 The point is argued in, e.g., Collins (2009).

22 Nor are they going to use it to justify mass unemployment. It seems likely that ex-chemist Margaret Thatcher was over-enamoured of the science of economics. Western democracies' governing bodies are notoriously short on scientists and we need many more of them. But let us hope that no scientist who believes that science can solve difficult problems of government is ever put in charge. If scientists are to be put in a decision-making position, they should be drawn from 'The Owls' – see chapter 3.

23 The nearest thing to it that we know of is the philosophers' idea of 'virtue epistemology'. But virtue epistemology appears to be based on the virtue of individuals rather than the nature of collective beliefs or virtuous aspirations (http://plato.stanford.edu/entries/epistemology-virtue). Dupré (1995) comes closest to what is argued here; it also invokes Wittgenstein in support of science being best described in terms of family resemblance (see chapter 1 below). Dupré says: 'I suggest that we try to replace the kind of epistemology that unites pure descriptivism and scientistic apologetics with something more like a virtue epistemology. There are many possible and actual such virtues: sensitivity to empirical fact, plausible background assumptions, coherence with other things we know, exposure to criticism from the widest variety of sources, and no doubt others.... it will hardly be difficult to demonstrate in such terms the greater

credibility earned by the subtle arguments and herculean marshaling of empirical facts of a Darwin, followed by a century and more of further empirical research and theoretical criticism, than that due to the attempt to ground historical matters of fact on the oracular interpretations of an ancient book of often unknown or dubious provenance' (p. 243). Incidentally, the authors of this book were ignorant of virtue epistemology until quite recently and it played no part in the development of elective modernism. Of course, it has long been known that science cannot go forward unless its proponents are virtuous, because the transmission of scientific facts relies on the trustworthiness of those doing experiments and reporting their results. Shapin (1994) traces the sources of belief in the personal integrity of scientists. But this, of course, is an argument *from* efficiency, not from the absolute goodness of the values themselves.

24 The precursor of this position is found in Collins (2001).

25 It is like Collins's (1985/1992) defence of replication as central to scientific method, even though it does not work as a clean and clear test as it was once thought to work. We will also argue later on that one particular virtue of scientific expertise is that it should be attentive to what it does not know as well as to what it does know.

26 For example, the 'Climategate' email case, in which emails sent between members of a scientific research group were hacked and made public, led to allegations of scientific fraud or malpractice, but only those with a detailed understanding of the science involved would be able to say whether or not the emails, or the practices they described, were inappropriate.

27 See Shapin (1979). This example is used in the Third Wave paper to make the same point; see Collins and Evans (2002).

28 These are paragraphs 66 and 67 of *Philosophical Investigations* (Wittgenstein, 1953), which has become a standard source for sociologists of scientific knowledge. Bloor (e.g., 1983) considers that in his later work Wittgenstein was as much a sociologist as a philosopher; certainly that is how sociologists of scientific knowledge treat him. See also Winch (1958).

29 There are competing interpretations of the notion of form of life. We make no attempt here to settle the question of what Wittgenstein really had in mind and adopt an interpretation that aligns with other sociological concepts such as 'taken-for-granted

reality' (Schutz, 1964), 'social collectivity' (Durkheim, 1915), 'paradigm' (Kuhn, 1962), culture (Kluckhohn, 1962; Geertz, 1973) and subculture (Yinger, 1982).

30 The fractal metaphor can be found in Collins and Kusch (1998). It is developed further in Collins (2011), which also introduces the term 'practice language', and Collins and Evans (2015b), which explains its relevance for social science research methods.

31 The founding father of this approach is Emile Durkheim. The point is developed in Collins (2011).

32 For the introduction of these concepts, see Collins and Kusch (1998).

33 In 2005, the philosopher Steve Fuller fell into the trap of using family resemblance too loosely while defending the teaching of intelligent design in schools in a US court case (*Kitzmiller* v. *Dover Area School District*). He said that, since religion was central to Newton's ideas, and that since the science of Newton and that of today bear a family resemblance, there is no conflict between evolution and a religiously motivated theory like intelligent design. Fuller seems not to have noticed that the argument applies equally to alchemy. Hasok Chang is another historian/philosopher whose view is that certain theories were never fully defeated by rational argument and that they should be revived. Chang does not understand that the replacement of one scientific theory by another is a social process. For a devastating critique, see Kusch (2015).

34 Robert Millikan's ignoring of certain results as recorded in his notebooks while he was establishing the unitary charge on the electron (Holton, 1978) is a useful example.

35 For 'strings', see Collins (2010).

36 None of the remarks made here about bodies of academics apply to all the academics in a discipline. We are trying to capture something about the broad characteristics of the thinking of an epoch – something entirely normal for historians of ideas, even though individuals and even minority groups may not fit. We attempted the same in characterizing science studies as made up of 'three waves' (Collins and Evans, 2002, 2007). When this kind of broad-brush approach is applied to contemporary thinkers, exceptions are more salient but, even though some violence must be committed, it still seems appropriate for sociologists of knowledge to try to describe

the 'ways of thinking' of their contemporaries, or near contemporaries, which are part of their 'ways of being in the world'. In the case of the characteristics of philosophers, it may be that our generalizations apply a little less to those who no longer embrace what we called 'wave one of science studies'. Since those days, philosophy has moved on and taken on board some of what was resisted in the early days of the 'sociological turn' in science studies. See, for example, the 'mission statement' of the Society for the Philosophy of Science in Practice: www.philosophy-science-practice.org/en/mission-statement.

37 The origins of the sentiment expressed in this paragraph go back to at least 1982 and emerge from consideration of how social scientists and natural scientists actually go about their work. The basic idea is that, to do their work well, both scientists and social scientists must act according to a set of aspirations that may not match what can actually be achieved; this, therefore, demands some compartmentalization.

38 Popper's *The Logic of Scientific Discovery* was published in German in 1934 and in English in 1959. Lakatos's ideas are most easily accessed in his 'Falsification and the Methodology of Scientific Research Programmes' (1970).

39 It is also unaffected by the fact that, occasionally, scientific observations are unique. Collins (1985, 1992) shows that, even in more ordinary cases, replication is impossible without making judgements that fall outside of the 'canonical view' of science, but that this does not mean that the idea of replication and the intention to replicate are not central to science. To bring the matter right up to date, consider the worries about single events and the relief at seeing a second gravitational wave that are reported in the story of the first detection of gravitational waves (Collins, 2017).

40 This is also how the Wave Two critique that makes science look more ordinary starts to build. If observation and falsification are practical problems, then their solution may look similar to those of other practical problems found in other domains. If this can be done, then science no longer looks unique and so loses its claim to special status.

41 The question of the way we cut up the world goes back at least as far as Plato. An attempt to research the means by which new ways

of cutting up the world are established is made by Collins (1985, 1992). This is a study of how scientists try to use experiments to establish the existence of novel objects. Unsurprisingly, experiments are never decisive but, nevertheless, groups of scientists still establish new ways of cutting up the world.

42 This is not to say that one might not reject the professed observations of one who is known to lack integrity or who is known to have deficient sensory organs or observational skills. One can say that 'in the normal way' one will prefer the word on matters observational of one who has observed over one who has not.

43 This is true even of Wave Two STS where the power of its critique comes from its careful *observation* of what scientists do when they are making truth.

44 The strain on the imagination is slightly lifted because the point has concrete implications in the way we should think about intelligent design – which is an unfalsifiable proposition. We should ask those who advocate intelligent design whether they would sacrifice falsifiability as a condition for the acceptance of their theory.

45 The counter-norms, which show that it is possible to argue that actions informed by principles that are diametrically opposed to those identified by Merton could still be seen as legitimate parts of science, were documented by Mitroff (1974).

46 The article from which this quote is drawn was originally published as Merton (1942). The page reference is from Merton (1979:270).

47 For Merton's tendency to cover all bases, see Shapin (1988).

48 Later on, when it had been revealed that science could be very different from the model imagined by Merton, he was even ready (1976), to countenance the notion of 'counter-norms'.

49 Note that, though elective modernism is not damaged by exceptional action tokens, Merton's scheme is, even though it is a sociological scheme. His scheme is justified by the efficacy of the values, so if the values are not essential to efficacy there is no justification for them.

50 See Collins (2017). Note that an ethic of communism is perfectly consistent with protecting participants' rights and commercial confidentiality – it is precisely because there is an expectation of sharing that these exceptions need to be justified by reference to other responsibilities. If secrecy was the norm, no such rules would

be needed as there would be no expectation of co-operation in the first place.

51 It is perfectly proper, of course, that political choices come into the setting of priorities as will be discussed at greater length in chapter 3.

52 For earlier arguments along these lines, see Martin, Richards and Scott (1991); Collins (1996).

53 At this point, what is being argued here comes much closer to 'virtue epistemology' since we are talking of the virtues of individuals. See also Shapin, (1994).

54 There are fine points here about when one action is embedded in another. Thus, the Manhattan Project was science embedded in the action of war-making. In the case of Fleck, it could be argued that he himself was embedding a scientific action – it requires scientific skill to make a bad vaccine – in a warlike action. But elective modernism is not concerned with outcomes: if Fleck's acts in creating a bad vaccine did not accord with the formative aspiration of science, then the form of life of science was not being exhibited. There is doubtless more philosophical work to be done here. For a discussion of how actions are embedded into other actions with conflicting intentions, see Collins and Kusch (1998).

55 See Collins and Evans (2007) for the origin of this idea.

56 And here we see how far we are from virtue epistemology. We should understand that the contemporary fashion for popularizing science is not all good. Scientists understand this: they understand that, say, the disproportionate amount of resources poured into planetary exploration and, perhaps, astronomy in general is a result of a certain public image of science, rather than any imperative growing from within science. In our terminology, this is pushing the LLI of science in the direction of the LLI of the arts.

57 George Orwell (1946) writes that obscurity bolsters fascism by enabling its operatives to plan and present violence, such as genocide, in mundane and euphemistic terms. Certain regions of the contemporary humanities and social sciences are notoriously obscure. Martha Nussbaum (1999) explains the way that this removes attention from the substance of the argument, instead creating a mysterious charisma and a band of followers for the arguer. Judith Butler is the particular target of Nussbaum's attack. Nussbaum also claims, however, that French postmodernism uses obscure language as a method of politi-

cal subversion, and Butler, who responds to Nussbaum (1999), also seems to justify obscure writing as a way of refusing to accept the power relations immanent in commonsense and its language (1999). In this book, we do not intend to bolster fascism nor are we trying to be subversive – indeed, like Orwell, we are trying to preserve what is valuable about Western societies, so it certainly behoves *us* to be clear.

58 See Collins, Bartlett and Reyes-Galindo (2016).

59 See Collins, Bartlett and Reyes-Galindo (2016), for discussion of the 'pathological individualism' that characterizes the 'fringe' of physics.

60 And, once more, intelligent design makes the point.

61 In 2004, just a couple of years after the 2002 publication of Collins and Evans's 'Third Wave' paper, one of the iconic figures belonging to the Second Wave of science studies, especially where those from the arts and humanities are concerned – Bruno Latour – expressed himself worried about global warming. In 'Why Has Critique Run Out of Steam?' he wrote: 'Was I wrong to participate in the invention of this field known as science studies? Is it enough to say that we did not really mean what we meant? Why does it burn my tongue to say that global warming is a fact whether you like it or not? Why can't I simply say that the argument is closed for good?' (2004:227).

62 Evans (1999).

63 For a discussion of the kinds of expertise that go into economic forecasting, and how forecasters are viewed by their clients, see Evans (2007).

64 This argument is taken from Nelson Goodman's *Languages of Art* where it is used to explain why we want to preserve the difference between fake art works and the real thing, even though we cannot, currently, see the difference between the two. The argument was rehearsed in Collins and Evans (2007).

65 Halfpenny (1982) sets out over a dozen meanings of positivism.

66 See also Collins, Bartlett and Reyes-Galindo (2016).

67 Collins (2007) makes the case that mathematics, though it is central to physics at the collective level, is not that important in physicists' day-to-day lives; Collins (1984) explains participant observation/ comprehension as a science.

68 The kind of ethnography or anthropology that prides itself on doing no more than describing the particulars of a specific situation is not science because the result is not general enough to be repeated.

69 For a discussion of humanities versus scientific approaches in social science, see Collins (2012). For more on 'subjective' social science methods, see Collins (2013a:ch. 16).

70 For collective tacit knowledge, see Collins (2010).

71 A more complete development of this point, under the heading of 'social cartesianism', will be found in Collins (2010).

72 See, e.g., Weinel (2010), who cites: Lane (1966); Brint (1990); Fischer (2000); Millstone (2009).

73 There were some instances of 'technocratic governments' being elected as a result of the financial crisis, e.g. that of Mario Monti in Italy (http://blogs.lse.ac.uk/europpblog/2013/06/11/the-rise-of-governments-led-by-technocrats-in-europe-illustrates-the-failure-of-mainstream-political-parties).

74 Krimsky (1984); Millstone (2009).

75 Krimsky (1984).

76 Wynne often writes in this vein. See, e.g., Wynne (1992a, 2003, 2008).

77 For examples, see Wynne (1982); Irwin (1995); Jasanoff (1995); Epstein (1996); Welsh (2000); Wickson and Wynne (2012).

78 This value–value distinction derives from the difference between intrinsic and extrinsic politics that was first set out in Collins and Evans (2002). Intrinsic politics refers to the residual political influences that must inevitably remain in any scientific endeavour no matter how rigorous and reflexive its practice. In other words, it is the politics that remains after all practical efforts have been made to eliminate it. In contrast, extrinsic politics refers to the explicit use of political concerns and interests to reach epistemic conclusions about data. SSK shows that intrinsic politics are inevitable; elective modernism argues that extrinsic politics must be eliminated even if intrinsic politics cannot. The formative aspiration of science is to remove all the politics, whether it can be accomplished or not: this is a central value and hence the value–value distinction.

79 As we said in *Rethinking Expertise* (Collins and Evans, 2007:8): 'Democracy cannot dominate every domain – that would destroy expertise – and expertise cannot dominate every domain – that would destroy democracy.'

80 This idea was developed by Martin Weinel (2010), where he called it the 'minimal default position'.

81 Wynne (2008) is an email exchange between various parties that took place after a talk given in Lancaster University by Robert Evans. We are grateful to Brian Wynne for allowing extracts from this private exchange of communications to be made public. For detailed analyses of the ways in which South African policy about AIDS can be understood in terms of that country's cultural history, see: Fassin (2007); Mahajan (2008).

82 Example references are: Nelkin (1971, 1975); Collingridge and Reeve (1986); Jasanoff (1990).

83 Wave Two analysts generally want to put the sandwich in a blender and say it is all one inseparable mixture of science and politics; we say much more is to be gained by keeping things as separate as possible. Elective modernism has its own set of formative intentions!

84 Brian Wynne quoted in a recent interview (Antonsen and Nilsen, 2013:37). If we take this statement at face value, then it is hard to see why Wynne is so opposed to the Third Wave / elective modernism initiative, as this is concerned with exactly the same issue: how to bring science into the democratic process without allowing it to determine policy outcomes.

85 See, e.g., Brush (1974); Collins (1982).

86 See Berger (1963) and Collins (2013a:ch. 16).

87 Owls have a variety of associations in myth and folklore. In the UK, owls are typically seen as wise, an association that may, in turn, date from the Ancient Greece where the 'Little Owl' was the messenger of Athene, the goddess of wisdom. In many other cultures, however, the owl has more negative connotations, often symbolizing sickness or death – that is unfortunate, and not what we intend.

88 Many scientists' lack of ability or willingness to reflect was on display during the 'science wars': the period when natural scientists violently attacked social scientists' analyses of science.

89 In the 'science wars', the philosopher-vultures were (and still are) among the most vicious of the predatory birds when it came to attacking the social scientists and, astonishingly, given the nature of philosophy proper, they are among the most vicious when it comes to attacking scientific heresies. Quite simply, though they are philosophers, they do not seem to understand philosophy – at least, they do not understand the philosophy of those of their colleagues who have rendered the logic of science puzzling.

90 For example, the magician 'The Amazing Randi' was deployed by *Nature* to discredit homeopathy, and Randi was treated by many scientists as an authority on the paranormal.

91 For a summary of the MMR controversy and a discussion of how parents may think about vaccination decisions, see Collins and Pinch (2005:esp. ch. 8).

92 See the discussion of core-sets in Collins (1985, 1992), and see Collins (2017:ch. 14) for more on the narrowness of expertises.

93 See the discussion of the sharp divide between primary source knowledge and interactional expertise found in Collins and Evans (2007) and Collins (2014a).

94 See the discussion of statistics below.

95 The MMR case was unusual in that there was no serious science supporting the anti-vaccination side.

96 To ensure a just distribution of food, Rawls suggested that those who divided up the rations should not choose their own portion – thus, he claimed, they would divide up the food as evenly as possible.

97 Quoted in Weinel (2007:752).

98 It also belongs to the topic of statistics which, it is becoming clear, is far trickier than has been thought. It goes without saying The Owls will also consider the extent to which scientists have acted according to the formative aspirations of science. Social scientists and investigative journalists are best at making this discrimination, since it is a central part of their tradition. Thus, journalists have discovered that the tobacco industry has paid certain scientists to create the appearance of controversy where there is none among the mainstream, while social scientists have found evidence that climate change 'sceptic' think-tanks have been funded by major oil companies, even though the relevant scientific community has reached consensus over the causes of climate change. All this will be taken into account in deciding on the nature of the consensus. Instances will be discussed in chapters 4 and 5.

99 Research on these topics is underway at Cardiff University, funded by ESRC grant RES/K006401/1 and British Academy Fellowship 'The Social Boundaries of Scientific Knowledge'. See the papers by Collins (2014b); Collins, Bartlett and Reyes-Galindo (2016); and Collins and Galindo (2016).

100 For an analysis of the case, see Collins, Bartlett and Reyes-Galindo (2016), citing Collins (2014c).

101 This will be most effective when the social scientists are familiar with the research community in question and so have the tacit knowledge needed to make good judgements. For an outline of a similar scheme, see Gorman and Schuurbiers (2013).

102 Should we wish to pursue the fable, these would constitute a 'College of Eagles'. The IPCC is a real College of Eagles; another example is the American JASONS.

103 Many of the points made in the next two paragraphs are discussed at greater length where we contrast the position of elective modernism with that of Heather Douglas.

104 To repeat, Collins and Evans (2002, 2007) distinguish between the intrinsic and the extrinsic politics of science. The intrinsic politics cannot be avoided but should be discounted whenever it is recognized; the extrinsic should never inform scientific activity.

105 Such as Jasanoff's (1990) 'scientific advisory committees', Guston's (2001) 'boundary organizations', Bijker, Bal and Hendriks's (2009) Gezondheidsraad, or Callon, Lascoumes and Barthe's (2010) 'hybrid forums'.

106 For science speaking to power, see Wildavsky (1979).

107 The quote is from the 'Third Wave of Science Studies' paper (Collins and Evans, 2002:240). 'Intellectual bankruptcy' for the analytic community does not mean moral bankruptcy or any kind of bankruptcy for the actors.

108 Fricker (2007).

109 See, e.g., Shapin (1994); Irwin (1995).

110 We say 'roughly' because we know some authors prefer not to be categorized in this way and we do not want to force them to accept our categorization.

111 See, e.g., Lynch and Cole (2005).

112 See, e.g., Wynne (1982).

113 See, e.g., Irwin (1995); Yearley (2000); Jenkins (2007); Callon, Lascoumes and Barthe (2010).

114 Wynne (1992a); Functowicz and Ravetz (1993).

115 Demortain (2013).

116 For a history of the regulations relating to seat belts and air bags in cars, see Wetmore (2015). For an example of how reported injuries

to children and women have led automobile researchers to consider a range of body shapes and sizes, see: www.ncbi.nlm.nih.gov/pmc/articles/PMC3400220.

117 Taleb (2010).

118 Callon, Lascoumes and Barthe (2010).

119 Oreskes and Conway (2010).

120 Hargreaves, Lewis and Speers (2003:2).

121 Hargreaves, Lewis and Speers (2003:42).

122 Laurent-Ledru, Thomson and Monsonego (2011). See also Sheldon (2009).

123 As we pointed out in chapter 1, where once science was used to defend democracy against fascism and communism, now science is positioned as being on the side of state control and regulation and as a threat to the individual liberties enjoyed by citizens in the democratic societies it once epitomized.

124 Quote is from Jasanof (2013:101).

125 For example: because more appointments, spread over a longer time, are needed; children are unprotected for longer; and there is a greater risk that they will not receive all the necessary injections.

126 Huxham and Sumner (1999), citing Nisbet and Fowler (1995).

127 Jasanoff is perhaps the scholar most famously associated with this approach, with *Designs on Nature* (Jasanoff 2007), which examines how different institutional forms in the UK, USA and Germany lead to very different laws on biotechnology, being rightly regarded as a classic in the field. The quote is from Jasanoff (2012:19).

128 See also the difference in the way Epstein's study of AIDS treatment activists is summarized in: Collins and Evans (2002) and Callon Lascoumes and Barthe (2010).

129 Cf. Collins (1996).

130 Turner (2003:131).

131 Turner (2003).

132 Brown (2009:203).

133 This section draws heavily on Jean Goodwin's paper presented at SEESHOP 6, held in June 2012 in Cardiff. We thank her for introducing us to the work of Walter Lippmann.

134 As we explain in Collins and Evans (2007), in such circumstances citizens instead make social judgements about who to trust. See also Evans (2011).

135 From Lippmann (1927).
136 Cf. Giddens (1990).
137 This is a rather harsh reading of his work. For more sympathetic readings, see: Jansen (2008, 2009); Schudson (2008).
138 For an account that discusses both Dewey and Lippmann, see Marres (2005).
139 See, e.g., Irwin (1995); Epstein (1996); Ottinger (2013).
140 Dewey (1954:155).
141 Schudson (2008).
142 A confusion that is arguably repeated in the debates following the publication of Collins and Evans (2002), where the idea of the problem of extension is seen as an attack on democracy.
143 The NIMBY problem is one example of this. Another, which doesn't rely on geography to define the engaged public is the GM Nation? debate in the UK. Comparisons of quantitative data collected from participants in open public meetings and from a representative sample of the population show that the active participants (i.e. the self-selected sample who went to public meetings) were significantly more sceptical than the public overall. See, e.g., Pidgeon et al. (2005); Horlick-Jones et al. (2007).
144 This also means that the political sphere is relatively autonomous in Rawls's theory as it does not depend directly on the support of any particular social group but arises independently and for different reasons.
145 To represent this graphically, you would have to imagine figures 3.1 and 3.2 re-drawn to include a number of the shaded ovals that overlapped to some extent with democratic values but did not encompass them completely. These could be things like Christian values, capitalist values, feminist values and so on.
146 Furthermore, whilst we assume that this political phase is characterized by democratic principles, these are not specified in any detail, meaning that that the political phase could, in principle, be operationalized in different ways. This seems to be consistent with Rawls's view that it makes more sense to talk of a family of political liberalisms than to insist on a single variant. See, e.g., Collins and Evans (2002, 2007); Evans and Plows (2007).
147 In practice, Rawls suggests that the restriction might be relaxed, subject to the provision that any justification expressed in terms of

a comprehensive doctrine could be translated into a justification expressed in terms of public reason.

148 Wenar (2013). For Rawls, the doctrine of public reason applies most strongly to fundamental political issues and to the statements of those holding public office. Outside these settings and in relation to other topics, citizens are free to reason however they choose, although Rawls does argue that, when acting in ways that relate to public office (e.g., running for an election or even just voting) then citizens should refer to public reason when taking part in these activities.

149 There is some resonance with civic epistemology here too, as, e.g., it is entirely possible that what counts as public reason – evidentiary standards and commonsense beliefs – will vary from place to place. Again, this seems consistent with Rawls, who elsewhere notes that even the fundamental values of fairness, equality and justice can be interpreted differently in different places, giving rise to many different versions of political liberalism, of which his is just one.

150 See Durant (2011).

151 Cohen (1999:399).

152 That is to say, whilst elective modernism would obviously encourage citizens to seek the best possible expert advice and act upon it in their private lives, there is no way in which this could or even should be enforced. All elective modernism says is that citizens should not claim to be specialist, technical experts when they are not. Apart from that, citizens are free to follow their own comprehensive doctrines and, for some, this will include choosing to ignore their doctors, avoid certain foods and/or believe in astrology.

153 In this sense, Habermas does differ from Rawls. Communicative action presumes a greater degree of shared understanding and agreement than Rawls's political liberalism, in which only the fundamental values agreed from behind the veil of ignorance are shared.

154 Habermas (1996:485).

155 This twin-track approach clearly has some structural similarities with the political liberalism of Rawls, most notably the distinction between the lifeworld (or background culture as Rawls would call it) and the formal political system. There are also some differences, particularly in how the two elements relate to each other. Rawls is basically silent about the connection, defining the political system as

relatively autonomous and hence separate from the comprehensive doctrines of everyday life. In contrast, the link between the two is an absolutely central issue for Habermas, as it is the correspondence between the communicative action of the lifeworld and the outcome of the political system that provides policy-makers with their legitimacy. This is why Habermas's theory is equivalent to a comprehensive doctrine for Rawls – it is a totalizing 'theory of everything' that explains how social relations should be ordered.

156 The term 'pragmatistic' is used in Habermas, *The Scientisation of Politics*. See Habermas (1970:66).

157 Remember that, in the case of Rawls, elective modernism is, effectively, a component of public reason.

158 Douglas (2009:45).

159 These standards can and do change over time. Research that was once seen as ethical can later be seen as unethical, and vice versa. Judgements about what is ethically acceptable are therefore relative to the cultural milieu of the host society and not an intrinsic part of science itself.

160 Douglas (2009:113–14).

161 Collins and Evans (2002, 2007).

162 She makes a similar point in 'Rejecting the Ideal of Value-Free Science', which appears in Douglas's (2007) collection, where she writes: 'The responsibility to consider the social and ethical consequences of one's actions and potential error cannot be sloughed off by scientists to someone else, without a severe loss of autonomy in research' (p. 130).

163 Douglas (2009:84).

164 Cf. Longino (1990); Harding (2006).

165 Douglas (2009:172–3).

166 Douglas (2009:163).

167 Furthermore, certain kinds of scientific brilliance or creativity seem to go with an absence of empathy with other citizens.

168 The statistical standard for a publication is 5 standard deviations in physics, and is occasionally found wanting in spite of growing from 3 in the 1960s as a result of experience; see Franklin (2013). Collins (2013a:ch. 5) shows why the meaning of even this rigorous standard is less clear-cut than it appears. Ioannidis (2005) argues that around half 'two standard deviation' results – the standard in most

other sciences – are unreliable. The Owls will have to understand the relationship between random and systematic error, trials factors, the file-drawer effect, pathological distributions in the parent population, and so on. Eagles in many sciences have shown that they cannot be trusted to handle statistics in the face of publication pressures but, to be fair, the problem is a hard one.

169 See Collins (2013a:ch. 5) for a discussion of these problems.

170 See Ioannidis (2005).

171 The argument is pre-figured in Selinger, Thompson and Collins (2011). In this paper Thompson argues, Douglas-like, that scientists who inveigh against genetically modified crops are acting immorally since the better nutrition and vitamin balance provided by such crops could save huge numbers of lives in the developing world and these lives are now being lost. Collins argues that if the anti-GMO scientists sincerely believe that the crops are dangerous, it is their duty to argue as hard as they can against their introduction. It is the job of politicians to decide where the balance of well-being lies. Under elective modernism, it would be the job of The Owls to decide on the balance of the science, and the job of the politicians to decide on the balance of well-being.

172 An example of the way this kind of thinking might go is Collins, Ginsparg and Reyes-Galindo (2016).

173 http://ndpr.nd.edu/news/29284–science-in-a-democratic-society.

174 http://ndpr.nd.edu/news/29284–science-in-a-democratic-society.

175 Kitcher (2012:33–4).

176 Brown (2013).

177 Although he does not reference it directly, Kitcher presumably has the 'Climategate' controversy caused by leaked emails in mind here. That said, it is not clear how this kind of response would really improve things. Indeed, it could be argued that the correct response to Climategate is more transparency, in the sense of a more open acknowledgement of the difficulties involved in doing cutting-edge science.

178 For an introductory overview, see Held (2006).

179 Durant (2011).

180 See, e.g., Mansbridge et al. (2012).

181 For a critical reflection on the value of these activities, see Rayner (2003).

182 The term comes from Callon, Lascoumes and Barthe (2010).

183 Rowe and Frewer (2004, 2005).

184 House of Lords (2000). Evans and Plows (2007) cite the following: Royal Commission for Environmental Pollution (1998); House of Lords (2000); Gerold and Liberatore (2001); Hargreaves and Ferguson (2001); Parliamentary Office of Science and Technology (2001); Office of Science and Technology (2002); Wilsden and Willis (2004); Council for Science and Technology (2005). See also Hagendijk (2004).

185 For discussions of the principles that underpin citizen juries and related methods, see, e.g., Grundahl (1995); Glasner and Dunkerley (1999); Guston (1999); Wakeford (2002); Evans and Kotchetkova (2009).

186 For the Netherlands, see Bijker, Bal and Hendriks (2009), Den Butter and Ten Wolde (2011); for Scandinavia, see Andersen and Jæger (1999), Zurita (2006), Nielsen, Lassen and Sandøe (2007). For the other countries, see Guston (1999); Purdue (1999); Einsiedel, Jelsøe and Breck (2001); Goven (2003); Nishizawa (2005); Seifert (2006); Dryzek and Tucker (2008). For a more complete listing, at least at the time of publication, see: Rowe and Frewer (2004, 2005).

187 Evans and Plows (2007); Evans (2011).

188 Schot and Rip (1997:251).

189 Schot (1998). See also: Rip, Misa and Schot (1995); Grin, van de Graaf and Hoppe (1997).

190 Schot (1998).

191 Schot (1998).

192 Rip and te Kulve (2008).

193 Kirsten et al. (2007); Retèl et al. (2009).

194 This is the locus of legitimate interpretation. See Collins and Evans (2002, 2007).

195 For more information, see: www.birds.cornell.edu.

196 For more information, see: http://setiathome.berkeley.edu.

197 This definition is closer to that used by Irwin (1995).

198 Brown (1987); Wynne (1992a); Epstein (1996); Popay and Williams (1996).

199 This is Natalie Jeremijenko's 'feral dog' project. For more information, see: www.nyu.edu/projects/xdesign.

200 Irwin (2001); Horlick-Jones et al. (2007).

201 The GM Nation? final report was published as Department of Trade
    and Industry (2003). The independent evaluation of the public
    debate is Horlick-Jones et al. (2007). Pidgeon et al. (2005) provides
    an analysis of the survey element in the GM Nation? debate.
202 See, e.g., Irwin (2001).
203 For overview of public participation in Europe from an STS per-
    spective, see: Hagendijk and Irwin (2006); Horst et al. (2007).
204 Evans and Plows (2007).
205 www.generationscotland.org/index.php?option=com_content&vie
    w=article&id=52&Itemid=124.
206 Haddow, Cunningham-Burley and Murray (2011).
207 This section is deeply indebted to the work of Martin Weinel
    (2010).
208 See, e.g.: Lane (1966); Ezrahi (1971); Krimsky (1984); Collingridge
    and Reeve (1986); Wynne (1989, 1992b); Fischer (1990); Millstone
    (2009).
209 Kantrowitz (1967).
210 Kantrowitz (1967:763).
211 Pielke (2008).
212 Irwin (2001:4). Evans's own observations of the reception given
    to the Nano-Jury report by the Chair of the House of Commons
    Committee with responsibility for nanotechnology and innovation
    tell a similar story.
213 Wynne (1982); Irwin (2001).
214 The classic Merton reference is his (1942) 'Science and Technology
    in a Democratic Order', which was later re-titled 'The Normative
    Structure of Science' (1979). Karl Popper's *The Open Society and its
    Enemies* (1945) and *The Poverty of Historicism* (1957) use an analysis
    of science to show that a science of history is impossible. It followed
    that totalitarian regimes based on theories of history could not sci-
    entifically justify mass reorganization of society. Elective modernism
    differs from Merton in respect of some of the norms, and more obvi-
    ously in the fact that it does not try to base the norms on the efficacy
    of science. It differs from Popper in that science is taken to be a
    form of life rather than a matter of 'logic'. As with Merton, then, the
    norms, including the norm associated with falsifiability, come before
    the science, rather than after it.
215 It is hard to see how a purely Wave Two model of science could

argue against the purchase of scientists for commercial advantage – it would just be another interest-driven way of doing science.

216 As explained, there are those – e.g. Stephen Turner – who argue that expert knowledge poses a particular challenge to liberal democracies, in that expertise, by its specialist nature, is inherently resistant to public control. We have discussed this in chapter 4.

217 *Nota bene* that elective modernism deals with the notion of scientific consensus or disagreement about the fact of the matter. Elective modernism, at least at this stage, has only limited things to say about the direction of scientific research. The position of elective modernism on such 'upstream' matters has been discussed in greater detail in chapter 3, under the heading of the 'sandwich model' of science and society.

218 This is what is referred to as 'Scientism4' in Collins and Evans's *Rethinking Expertise* (2007:10), where scientism has the following definitions. **Scientism1**: an over-pedantic cleaving to some canonical model of scientific method or reasoning. **Scientism2**: scientific fundamentalism: a zealot-like view that the only sound answer to any question is to be found in science or scientific method. **Scientism3**: the view that narrowly framed 'propositional questions' posed by scientific experts are the only legitimate way to approach a debate concerned with science and technology in the public domain; this goes along with blindness to the political embeddedness of such questions. **Scientism4**: the view that science should be treated not just as a resource, but as a central element of our culture. Scientism1–Scientism3 are rejected under elective modernism.

219 See Collins (2014a) for an accessible treatment of the changing 'zeitgeist'.

220 There are parallels with the work of the US Office of Technology Assessment (OTA) that was created in 1972 and then abolished in 1995. The role of the OTA was to supply Congress with information on emerging technologies based on the best scientific and technical consensus the OTA could find and their reports were also made available to the public. For more on the history of the OTA see: Sadowski (2015).

References

Agriculture and Environment Biotechnology Commission. 2001. *Crops on Trial: A Report by the AEBC*. London: Agriculture and Environment Biotechnology Commission.

Andersen, I. E. and Jæger, B. 1999. Scenario Workshops and Consensus Conferences: Towards More Democratic Decision-making. *Science and Public Policy*, 26(5): 331–40.

Antonsen, Marie and Nilsen, Rita Elmkvist. 2013. Strife of Brian: Science and Reflexive Reason as a Public Project. An Interview with Brian Wynne. *Nordic Journal of Science and Technology Studies*, 1(1): 31–40.

Berger, P. L. 1963. *Invitation to Sociology*. Garden City: Anchor Books.

Bijker, Wiebe E., Bal, Roland and Hendriks, Ruud. 2009. *The Paradox of Scientific Authority*. Cambridge, MA: MIT Press.

Bloor, D. 1983. *Wittgenstein: A Social Theory of Knowledge*. London: Macmillan.

Brint, S., 1990. Rethinking the Policy Influence of Experts: From General Characterizations to Analysis of Variation. *Sociological Forum*, 5(3): 361–85.

Brown, Mark B. 2009. *Science in Democracy: Expertise, Institutions, and Representation*. Cambridge, MA: MIT Press.

Brown, Mark B. 2013. Review of Philip Kitcher, *Science in a Democratic Society*. *Minerva*, 51: 389–97.

Brown, P. 1987. Popular Epidemiology: Community Response to Toxic Waste-induced Disease in Woburn, Massachusetts. *Science, Technology, & Human Values*, 12(3/4): 78–85.

Brush, Stephen G. 1974. Should the History of Science Be Rated X? *Science*, 183 (4130): 1164–72.

Butler, Judith. 1999. 'Bad Writer' Bites Back. *New York Times* op-ed, 20 March.

Callon, Michel, Lascoumes, Pierre and Barthe, Yannick. 2010. *Acting in an Uncertain World: An Essay on Technical Democracy*. Cambridge, MA: MIT Press.

Carr, E. Summerson. 2010. Enactments of Expertise. *Annual Review of Anthropology*, 39(1): 17–32. doi:10.1146/annurev.anthro.012809. 104948.

Cohen, J. 1999. Reflections on Habermas on Democracy. *Ratio Juris*, 12(4): 385–416.

Collingridge, D. and Reeve, C. 1986. *Science Speaks to Power: The Role of Experts in Policy Making*. New York: St Martin's Press.

Collins, Harry M. 1975. The Seven Sexes: A Study in the Sociology of a Phenomenon, or the Replication of Experiments in Physics. *Sociology*, 9(2): 205–24.

Collins, Harry M. 1982. Special Relativism: The Natural Attitude. *Social Studies of Science*, 12: 139–43.

Collins, Harry M. 1984. Concepts and Methods of Participatory Fieldwork, in C. Bell and H. Roberts (eds.), *Social Researching*. Henley-on-Thames: Routledge, 54–69.

Collins, Harry M. 1985. *Changing Order: Replication and Induction in Scientific Practice*. Beverley Hills and London: Sage.

Collins, Harry M. 1992. *Changing Order: Replication and Induction in Scientific Practice*. 2nd edition. Chicago: University of Chicago Press.

Collins, Harry M. 1996. In Praise of Futile Gestures: How Scientific is the Sociology of Scientific Knowledge?, in *The Politics of SSK: Neutrality, Commitment and Beyond*: Special Issue of *Social Studies of Science*, 26(2): 229–44.

Collins, Harry M. 2001. Crown Jewels and Rough Diamonds: The Source of Science's Authority, in Jay Labinger and Harry Collins (eds.),

The One Culture? A Conversation about Science. Chicago: University of Chicago Press, 255–60.

Collins, Harry M. 2004a. *Gravity's Shadow: The Search for Gravitational Waves.* Chicago: University of Chicago Press.

Collins, Harry M. 2004b. Interactional Expertise as a Third Kind of Knowledge. *Phenomenology and the Cognitive Sciences*, 3(2): 125–43.

Collins, Harry M. 2007. Mathematical Understanding and the Physical Sciences, in Harry M. Collins (ed.), *Case Studies of Expertise and Experience*: Special Issue of *Studies in History and Philosophy of Science*, 38(4): 667–85.

Collins, Harry M. 2009. We cannot live by scepticism alone. *Nature*, 458(March): 30–1.

Collins, Harry M. 2010. *Tacit and Explicit Knowledge.* Chicago: University of Chicago Press.

Collins, Harry M. 2011. Language and Practice. *Social Studies of Science*, 41(2): 271–300.

Collins, Harry M. 2012. Performances and Arguments. *Metascience*, 21(2): 409–18.

Collins, Harry M. 2013a. *Gravity's Ghost and Big Dog: Scientific Discovery and Social Analysis in the Twenty-First Century.* Enlarged edition. Chicago: University of Chicago Press.

Collins, Harry M. 2013b. Three Dimensions of Expertise. *Phenomenology and the Cognitive Sciences*, 12(2): 253–73.

Collins, Harry M. 2014a. *Are We All Scientific Experts Now?* Cambridge: Polity.

Collins, Harry M. 2014b. Rejecting Knowledge Claims Inside and Outside Science. *Social Studies of Science*, 44(5): 722–35. doi:10.1177/0306312714536011.

Collins, Harry M. 2017. *Gravity's Kiss: The Detection of Gravitational Waves.* Chicago: University of Chicago Press.

Collins, Harry, Bartlett, Andrew and Reyes-Galindo, Luis. 2016. *The Ecology of Fringe Science and its Bearing on Policy.* http://arxiv.org/abs/1606.05786.

Collins, Harry M. and Evans, Robert. 2002. The Third Wave of Science Studies: Studies of Expertise and Experience. *Social Studies of Sciences*, 32(2): 235–96.

Collins, Harry M. and Evans, Robert. 2003. King Canute Meets the Beach Boys: Responses to the Third Wave. *Social Studies of Science*, 33(3): 435–52.

Collins, Harry and Evans, Robert. 2007. *Rethinking Expertise*. Chicago: University of Chicago Press.

Collins, Harry M. and Evans, Robert. 2014. Quantifying the Tacit: The Imitation Game and Social Fluency. *Sociology*, 48(1): 3–19.

Collins, Harry M. and Evans, Robert. 2015a. Expertise Revisited, Part I – Interactional Expertise. *Studies in History and Philosophy of Science Part A*, 54(December): 113–23. doi:10.1016/j.shpsa.2015.07.004.

Collins, Harry M., and Evans, Robert. 2015b. Probes, Surveys, and the Ontology of the Social. *Journal of Mixed Methods Research* (December). doi:10.1177/1558689815619825.

Collins, Harry M., Evans, Robert, Ribeiro, Rodrigo and Hall, Martin. 2006. Experiments with Interactional Expertise. *Studies in History and Philosophy of Science Part A*, 37(4): 656–74.

Collins, Harry M., Evans, Robert and Weinel, Martin. 2016. Expertise Revisited, Part II: Contributory Expertise. *Studies in History and Philosophy of Science Part A*, 56(April): 103–10. doi:10.1016/j.shpsa.2015.07.003.

Collins, Harry M., Evans, Robert, Weinel, Martin, Lyttleton-Smith, Jennifer, Bartlett, Andrew and Hall, Martin. 2015. The Imitation Game and the Nature of Mixed Methods. *Journal of Mixed Methods Research* (December). doi:10.1177/1558689815619824.

Collins, Harry M., Ginsparg, Paul and Reyes-Galindo, Luis. 2016. A Note Concerning Primary Source Knowledge. *Journal of the Association for Information Science and Technology*. http://arxiv.org/abs/1605.07228.

Collins, Harry M. and Kusch, Martin. 1998. *The Shape of Actions: What Humans and Machines Can Do*. Cambridge, MA: MIT Press.

Collins, Harry M. and Pinch, Trevor. 2005. *Dr Golem: How to Think about Medicine*. Chicago: University of Chicago Press.

Collins, Harry M., Weinel, Martin and Evans, Robert. 2010. The Politics and Policy of the Third Wave: New Technologies and Society. *Critical Policy Studies*, 4(2): 185–201.

Collins, Harry M., Weinel, Martin and Evans, Robert. 2011. Object and Shadow: Responses to the CPS Critiques of Collins, Weinel and Evans' 'Politics and Policy of the Third Wave'. *Critical Policy Studies*, 5(3): 340–8.

Council for Science and Technology (CST). 2005. *Policy Through Dialogue*. London: CST. www.cst.org.uk/reports.

Darling, Karen M. 2003. Motivational Realism: The Natural Classification for Pierre Duhem. *Philosophy of Science*, 70(December): 1125–36.

Demortain, D. 2013. Regulatory Toxicology in Controversy. *Science, Technology & Human Values*, 38(6): 727–48.

Den Butter, F. A. and Ten Wolde, S. 2011. The Institutional Economics of Stakeholder Consultation: Reducing Implementations Costs through 'Matching Zones'. Tinbergen Institute Discussion paper TI 2011–162/3. Amsterdam and Rotterdam: Tinbergen Institute.

Department of Trade and Industry (DTI). 2003. *GM Nation? The Findings of the Public Debate*. London: DTI.

Dewey, J. 1954. *The Public and its Problems*. Athens, OH: Swallow Press. (Original work published 1927)

Douglas, Heather. 2007. Rejecting the Ideal of Value-Free Science, in Harold Kincaid, John Dupré, and Alison Wylie (eds.), *Value-Free Science? Ideals and Illusions*. Oxford and New York: Oxford University Press, 120–39.

Douglas, Heather. 2009. *Science, Policy and the Value-Free Ideal*. Pittsburgh: University of Pittsburgh Press.

Douma, Kirsten F. L., Kim Karsenberg, Marjan J. M. Hummel, Jolien M. Bueno-de-Mesquita and Wim H. van Harten. 2007. Methodology of Constructive Technology Assessment in Health Care. *International Journal of Technology Assessment in Health Care*, 23: 162–8. doi:10.1017/S0266462307070262.

Dryzek, J. S. and Tucker, A. 2008. Deliberative Innovation to Different Effect: Consensus Conferences in Denmark, France, and the United States. *Public Administration Review*, 68(5): 864–76.

Dupré, John. 1995. *The Disorder of Things*. Cambridge, MA: Harvard University Press.

Durant, Darrin. 2011. Models of Democracy in Social Studies of Science. *Social Studies of Science*, 41(5): 691–714.

Durkheim, Émile. 1915. *Elementary Forms of the Religious Life*. London: George Allen and Unwin.

Durkheim, Émile. 1958. *Professional Ethics and Civic Morals*. Glencoe: Free Press.

Einsiedel, E. F., Jelsøe, E. and Breck, T. (2001). Publics at the Technology Table: The Consensus Conference in Denmark, Canada, and Australia. *Public Understanding of Science*, 10(1): 83–98

Epstein, Steven. 1996. *Impure Science: AIDS, Activism, and the Politics of Knowledge.* Berkeley: University of California Press.

Epstein, Steven. 2011. Misguided Boundary Work in Studies of Expertise: Time to Return to the Evidence. *Critical Policy Studies* 5(3): 323–8.

Evans, Robert. 1999. *Macroeconomic Forecasting: A Sociological Appraisal.* London: Routledge.

Evans, Robert. 2007. Social Networks and Private Spaces in Economic Forecasting, in Harry Collins (ed.), *Case Studies of Expertise and Experience:* Special Issue of *Studies in History and Philosophy of Science,* 38(4): 686–97.

Evans, Robert. 2011. Collective Epistemology: The Intersection of Group Membership and Expertise, in Hans Bernhard Schmid, Daniel Sirtes and Marcel Weber (eds.), *Collective Epistemology.* Heusenstamm: Ontos Verlag, 177–202.

Evans, Robert and Crocker, Helen. 2013. The Imitation Game as a Method for Exploring Knowledge(s) of Chronic Illness. *Methodological Innovations Online,* 8(1): 34–52. doi:10.4256/mio.2013.003.

Evans, Robert and Collins, Harry M. 2010. Interactional Expertise and the Imitation Game, in M. E. Gorman (ed.), *Trading Zones and Interactional Expertise Creating New Kinds of Collaboration, Inside Technology.* Cambridge, MA: MIT Press, 53–70.

Evans, Robert and Kotchetkova, Inna. 2009. Qualitative Research and Deliberative Methods: Promise or Peril? *Qualitative Research,* 9(5): 625–43.

Evans, Robert and Plows, Alexandra. 2007. Listening Without Prejudice? Re-Discovering the Value of the Disinterested Citizen. *Social Studies of Science,* 37(6): 827–54.

Eyal, Gil. 2013. For a Sociology of Expertise: The Social Origins of the Autism Epidemic. *American Journal of Sociology,* 118(4): 863–907.

Ezrahi, Y. 1971. The Political Resources of American Science. *Science Studies,* 1(2): 117–33.

Fassin, Didier. 2007. *When Bodies Remember: Experiences and Politics of AIDS in South Africa.* Berkeley: University of California Press.

Fischer, Frank. 1990. *Technocracy and Politics of Expertise.* Newbury Park, London and New Delhi: Sage.

Fischer, Frank. 2000. *Citizens, Experts, and the Environment: The Politics of Local Knowledge.* Durham, NC: Duke University Press.

Fischer, Frank. 2009. *Democracy and Expertise: Reorienting Policy Inquiry*. Oxford: Oxford University Press.

Fischer, Frank. 2011. The 'Policy Turn' in the Third Wave: Return to the Fact–Value Dichotomy?' *Critical Policy Studies*, 5(3): 311–16.

Forsyth, Tim. 2011. Expertise Needs Transparency Not Blind Trust: A Deliberative Approach to Integrating Science and Social Participation. *Critical Policy Studies*, 5(3): 317–22.

Franklin, Allan. 2013. *Shifting Standards: Experiments in Particle Physics in the Twentieth Century*. Pittsburgh: University of Pittsburgh Press.

Fricker, Miranda. 2007. *Epistemic Injustice: Power and the Ethics of Knowing*. Oxford: Oxford University Press.

Functowicz, Silvio O. and Ravetz, Jerome R. 1993. Science in the Post-Normal Age. *Futures*, 25(7): 739–55.

Geertz, C. 1973. *The Interpretation of Cultures*. New York: Basic Books.

Gerold, Rainer and Liberatore, Angela. 2001. *Report of the Working Group 'Democratising Expertise and Establishing Scientific Reference Systems'*. Brussels: European Commission. http://europa.eu.int/comm/governance/areas/group2/report_en.pdf.

Giddens, Anthony. 1990. *The Consequences of Modernity*. Cambridge: Polity.

Glasner, P. and Dunkerley, D. 1999. The New Genetics, Public Involvement, and Citizens' Juries: A Welsh Case Study. *Health, Risk & Society*, 1(3): 313–24.

Goodman, Nelson. 1968. *Languages of Art: An Approach to a Theory of Symbols*. Indianapolis, IN: Bobbs-Merrill.

Gorman, Michael E. 2002. Levels of Expertise and Trading Zones: A Framework for Multidisciplinary Collaboration. *Social Studies of Science*, 32(5–6): 933–8.

Gorman, Michael E., and Schuurbiers, Daan. 2013. Convergence and Crossovers in Interdisciplinary Engagement with Science and Technology, in K. Konrad, C. Coenen, A. Dijkstra, C. Milburn and H. van Lente (eds.), *Shaping Emerging Technologies: Governance, Innovation, Discourse*. Berlin: Akademische Verlagsgesellschaft, 7–20.

Goven, J. 2003. Deploying the Consensus Conference in New Zealand: Democracy and De-problematization. *Public Understanding of Science*, 12(4): 423–40. www.ruhr-uni-bochum.de/kbe/Korean_Consensus.html.

Grin, J., van de Graaf, H. and Hoppe, R. 1997. *Technology Assessment through Interaction: A Guide*. The Hague: Rathenau Institute.

Grundahl, J. 1995. The Danish Consensus Conference Model, in S. Joss and J. Durant (eds.), *Public Participation in Science: The Role of Consensus Conferences in Europe*. London: Science Museum, 31–40. http://people. ucalgary.ca/~pubconf/Education/grundahl.htm.

Guston, David H. 1999. Evaluating the First U.S. Consensus Conference: The Impact of the Citizens' Panel on Telecommunications and the Future of Democracy. *Science, Technology & Human Values*, 24(4): 451–82.

Guston, David H. 2001. Boundary Organizations in Environmental Policy and Science: An Introduction. *Science, Technology, & Human Values*, 26(4): 399–408.

Habermas, Jurgen. 1970. *Toward a Rational Society: Student Protest, Science, and Politics*. Boston, MA: Beacon Press.

Habermas, Jurgen. 1996. *Between Facts and Norms*, trans. William Rehg. Cambridge, MA: MIT Press.

Haddow, G., Cunningham-Burley, S. and Murray, L. 2011. Can the Governance of a Population Genetic Data Bank Effect Recruitment? Evidence from the Public Consultation of Generation Scotland. *Public Understanding of Science*, 20(1): 117–29.

Hagendijk, R. P. 2004. The Public Understanding of Science and Public Participation in Regulated Worlds. *Minerva*, 42(1): 41–59.

Hagendijk, Rob and Irwin, Alan. 2006. Public Deliberation and Governance: Engaging with Science and Technology in Contemporary Europe. *Minerva*, 44(2): 167–84.

Halfpenny, Peter. 1982. *Positivism and Sociology*. London: George Allen and Unwin.

Hanlon, Gerard. 1999. *Lawyers, the State and the Market: Professionalism Revisited*. Basingstoke: Macmillan.

Harding, Sandra. 2006. *Science and Social Inequality: Feminist and Postcolonial Issues*. Urbana and Chicago: University of Illinois Press.

Hargreaves, Ian and Ferguson, Galit. 2001. *Who's Misunderstanding Whom? Bridging the Gulf of Understanding between the Public, the Media and Science*. Swindon, Wilts.: ESRC.

Hargreaves, Ian, Lewis, Justin and Speers, Tammy. 2003. *Towards a Better Map: Science, the Public and the Media*. Swindon, Wilts.: ESRC.

Held, D. 2006. *Models of Democracy*. 3rd edition. Cambridge: Polity.

Holton, Gerald. 1978. *The Scientific Imagination: Case Studies*. Cambridge: Cambridge University Press.

Horlick-Jones, Tom, Walls, John, Rowe, Gene, et al. 2007. *The GM Debate: Risk, Politics and Public Engagement*. London: Routledge.

Horst, Maja, Irwin, Alan, Healey, Peter and Hagendijk, Rob. 2007. European Scientific Governance in a Global Context: Resonances, Implications and Reflections. *IDS Bulletin*, 38(5): 6–20.

House of Lords. 2000. *Science and Society: Science and Technology Select Committee, Third Report*. London: HMSO. www.parliament.the-stationery-office.co.uk/pa/ld199900/ldselect/ldsctech/38/3801.htm.

Huxham, Mark and Sumner, David. 1999. Emotion, Science and Rationality: The Case of the Brent Spar. *Environmental Values*, 8(3): 349–68.

Ioannidis, J. P. A. 2005. Why Most Published Research Findings are False. *PLoS Medicine*, 2(8): e124, 696–701.

Irwin, Alan. 1995. *Citizen Science: A Study of People, Expertise and Sustainable Development*. London and New York: Routledge.

Irwin, Alan. 2001. Constructing the Scientific Citizen: Science and Democracy in the Biosciences. *Public Understanding of Science*, 10(1): 1–18.

Jansen, Sue Curry. 2008. Walter Lippmann, Straw Man of Communication Research. In David W. Park and Jefferson Pooley (eds.), *The History of Media and Communication Research: Contested Memories*. New York: Peter Lang Press, 1–23.

Jansen, Sue Curry. 2009. Phantom Conflict: Lippmann, Dewey, and the Fate of the Public in Modern Society. *Communication and Critical/Cultural Studies*, 6(3): 221–45.

Jasanoff, Sheila. 1990. *The Fifth Branch: Science Advisers as Policymakers*. Cambridge, MA: Harvard University Press.

Jasanoff, Sheila. 1995. *Science at the Bar: Law, Science, and Technology in America*. Cambridge, MA, and London: Harvard University Press.

Jasanoff, Sheila. 2003. Breaking the Waves in Science Studies: Comment on H. M. Collins and Robert Evans, 'The Third Wave of Science Studies'. *Social Studies of Science*, 33(3): 389–400.

Jasanoff, Sheila. 2007. *Designs on Nature: Science and Democracy in Europe and the United States*. Princeton, NJ: Princeton University Press. http://site.ebrary.com/id/10477123.

Jasanoff, Sheila. 2012. *Science and Public Reason*. London and New York: Routledge.

Jasanoff, Sheila. 2013. Fields and Fallows: A Political History of STS, in A. Barry and G. Born (eds.), *Interdisciplinarity: Reconfigurations of the Natural and Social Sciences*. London and New York: Routledge, 99–118.

Jenkins, L. D. 2007. Bycatch: Interactional Expertise, Dolphins and the US Tuna Fishery. *Studies in History and Philosophy of Science Part A*, 38(4): 698–712.

Jennings, Bruce. 2011. Poets of the Common Good: Experts, Citizens, Public Policy. *Critical Policy Studies*, 5(3): 334–9.

Kantrowitz, A. 1967. Proposal for an Institution for Scientific Judgment. *Science*, 156(12 May): 763–4.

Kitcher, Philip. 2001. *Science, Truth and Democracy*. Oxford: Oxford University Press.

Kitcher, Philip. 2012. *Science in a Democratic Society*. Amherst, NY: Prometheus Books.

Kluckhohn, R. (ed.). 1962. *Culture and Behavior: Collected Essays of Clyde Kluckhohn*. Glencoe, IL: Free Press of Glencoe.

Krimsky, S. 1984. Epistemic Considerations on the Value of Folk-wisdom in Science and Technology. *Policy Studies Review*, 3(2): 246–62.

Kuhn, T. S. 1962. *The Structure of Scientific Revolutions*. Chicago, IL: University of Chicago Press.

Kuhn, T. S. 1970. *The Structure of Scientific Revolutions*. 2nd edition. Chicago, IL: University of Chicago Press.

Kuhn, T. S. 1979. *The Essential Tension: Selected Studies in Scientific Tradition and Change*. Chicago, IL: University of Chicago Press.

Kusch, Martin. 2015. Scientific Pluralism and the Chemical Revolution. *Studies in History and Philosophy of Science*, 49: 69–79.

Lakatos, Imre. 1970. Falsification and the Methodology of Scientific Research Programmes, in I. Lakatos and A. Musgrave (eds.), *Criticism and the Growth of Knowledge*. Cambridge: Cambridge University Press, 91–196.

Lane, R. E. 1966. The Decline of Politics and Ideology in a Knowledgeable Society. *American Sociological Review*, 31(5): 649–62.

Latour, B. 2004. Why Has Critique Run Out of Steam? From Matters of Fact to Matters of Concern. *Critical Inquiry*, Winter: 225–48.

Laurent-Ledru, V., Thomson, A. and Monsonego, J. 2011. Civil Society: A Critical New Advocate for Vaccination in Europe. *Vaccine*, 29(4): 624–8.

Lippmann, Walter. 1927. *The Phantom Public.* http://www2.maxwell.syr. edu/plegal/history/lippmann.htm.

Longino, Helen. 1990. *Science as Social Knowledge.* Princeton, NJ: Princeton University Press.

Lynch, M. and Cole, S. 2005. Science and Technology Studies on Trial: Dilemmas of Expertise. *Social Studies of Science*, 35(2): 269–311.

Mahajan, M. 2008. The Politics of Public Health Emergencies: AIDS Epidemics in India and South Africa. Doctoral dissertation, Cornell University.

Mansbridge, Jane, Bohman, James, Chambers, Simone, et al. 2012. A Systemic Approach to Deliberative Democracy, in J. Parkinson and J. Mansbridge (eds.), *Deliberative Systems.* Cambridge: Cambridge University Press, 1–26.

Marres, Noortje S. 2005. No Issue, No Public: Democratic Deficits after the Displacement of Politics. Ph.D. thesis, Amsterdam School for Cultural Analysis (ASCA). http://dare.uva.nl/record/1/241881.

Martin, B., Richards, E. and Scott, P. 1991. Who's a Captive? Who's a Victim? Response to Collins's Method Talk. *Science Technology & Human Values*, 16(2): 252–5.

Merton, Robert. 1942. Science and Technology in a Democratic Order. *Journal of Legal and Political Sociology*, 1: 115–26.

Merton, Robert. 1976. *Sociological Ambivalence.* New York: Free Press

Merton, Robert. 1979. The Normative Structure of Science, in Robert K. Merton, *The Sociology of Science: Theoretical and Empirical Investigations.* Chicago: University of Chicago Press, 267–78.

Millstone, E. 2009. Science, Risk and Governance: Radical Rhetorics and the Realities of Reform in Food Safety Governance. *Research Policy*, 38(4): 624–36.

Mitroff, Ian I. 1974. Norms and Counter-Norms in a Select Group of the Apollo Moon Scientists: A Case Study of the Ambivalence of Scientists. *American Sociological Review*, 39(4): 579–95.

Nelkin, D. 1971. Scientists in an Environmental Controversy. *Social Studies*, 1: 245–61.

Nelkin, D. 1975. The Political Impact of Technical Expertise. *Social Studies of Science*, 5: 35–54.

Nielsen, A. P., Lassen, J. and Sandøe, P. 2007. Democracy at its Best? The Consensus Conference in a Cross-national Perspective. *Journal of Agricultural and Environmental Ethics*, 20(1): 13–35.

Nisbet, E. and Fowler, C. 1995. Is Metal Disposal Toxic to Deep Oceans? *Nature*, 375: 715.

Nishizawa, M. 2005. Citizen Deliberations on Science and Technology and their Social Environments: Case Study on the Japanese Consensus Conference on GM Crops. *Science and Public Policy*, 32(6): 479–89.

Nussbaum, Martha. 1999. The Professor of Parody. *The New Republic Online* (TheNewRepublic.com). www.tnr.com/index.mhtml 02.22.99.

Office of Science and Technology (OST). 2002. *The Government's Approach to Public Dialogue on Science and Technology*. London: OST. www.ost.gov.uk/society/public_dialogue.htm.

Oreskes, Naomi and Conway, Erik M. 2010. *Merchants of Doubt: How a Handful of Scientists Obscured the Truth on Issues from Tobacco Smoke to Global Warming*. New York: Bloomsbury Press.

Orwell, George. 1946. Politics and the English Language. www.orwell. ru/library/essays/politics/english/e_polit.

Ottinger, Gwen. 2013. *Refining Expertise: How Responsible Engineers Subvert Environmental Justice Challenges*. New York and London: New York University Press.

Owens, Susan. 2011. Three Thoughts on the Third Wave. *Critical Policy Studies*, 5(3): 329–33.

Parliamentary Office of Science and Technology (POST). 2001. *Open Channels: Public Dialogue in Science and Technology*. Report No. 153. London: Parliamentary Office of Science and Technology.

Parsons, Talcott. 1991. *The Social System*. New edn. London: Routledge.

Pielke, Roger A. 2007. *The Honest Broker: Making Sense of Science in Policy and Politics*. Cambridge; New York: Cambridge University Press.

Pidgeon, Nick F., Poortinga, Wouter, Rowe, Gene, Horlick-Jones, Tom, Walls, John and O'Riordan, Tim. 2005. Using Surveys in Public Participation Processes for Risk Decision Making: The Case of the 2003 British GM Nation? Public Debate. *Risk Analysis*, 25(2): 467–79. doi:10.1111/j.1539–6924.2005.00603.x.

Popay, J. and Williams, G. 1996. Public Health Research and Lay Knowledge. *Social Science & Medicine*, 42(5): 759–68.

Popper, Karl. 1945. *The Open Society and its Enemies*. London: Routledge and Kegan Paul.

Popper, Karl. 1957. *The Poverty of Historicism*. London: Routledge and Kegan Paul.

Popper, Karl. 1959. *The Logic of Scientific Discovery*. New York: Harper & Row.

Purdue, D. 1999. Experiments in the Governance of Biotechnology: A Case Study of the UK National Consensus Conference. *New Genetics and Society*, 18(1): 79–99.

Rayner, Steve. 2003. Democracy in the Age of Assessment: Reflections on the Roles of Expertise and Democracy in Public-Sector Decision Making. *Science and Public Policy*, 30:2 (June): 163–70.

Retèl, Valesca P., Bueno-de-Mesquita, Jolien M., Hummel, Marjan J. M., et al. 2009. Constructive Technology Assessment (CTA) as a Tool in Coverage with Evidence Development: The Case of the 70-gene Prognosis Signature for Breast Cancer Diagnostics. *International Journal of Technology Assessment in Health Care*, 25(1): 73–83.

Rip, Arie. 2003. Constructing Expertise: In a Third Wave of Science Studies? *Social Studies of Science*, 33(3): 419–34.

Rip, Arie and te Kulve, Haico. 2008. Constructive Technology Assessment and Sociotechnical Scenarios, in Erik Fisher, Cynthia Selin and Jameson M. Wetmore (eds.), *The Yearbook of Nanotechnology in Society*, Volume I: *Presenting Futures*. Berlin: Springer, 49–70.

Rip, Arie, Misa, Thomas J. and Schot, John (eds.). 1995. *Managing Technology in Society: The Approach of Constructive Technology Assessment*. London and New York: Pinter Publishers.

Rowe, Gene and Frewer, Lynn J. 2004. Evaluating Public Participation Exercises. *Science Technology & Human Values*, 29(4) (Autumn): 512–57.

Rowe, Gene and Frewer, Lynn J. 2005. A Typology of Public Engagement Mechanisms. *Science, Technology & Human Values*, 30(2) (Spring): 251–90.

Royal Commission for Environmental Pollution (RCEP). 1998. *21st Report: Setting Environmental Standards*, Cm 4053. London: HM Stationery Office.

Sadowski, Jathan. 2015. Office of Technology Assessment: History, Implementation, and Participatory Critique. *Technology in Society*, 42 (August): 9–20. doi:10.1016/j.techsoc.2015.01.002.

Schot, Johan. 1998. Constructive Technology Assessment Comes of Age: The Birth of a New Politics of Democracy. www.ifz.tu-graz.ac.at/sumacad/schot.pdf. Also published in A. Jamison (ed.), *Technology Policy Meets the Public*, PESTO papers II. Aalborg: Aalborg University, 207–32.

Schot, Johan and Rip, Arie. 1997. The Past and Future of Constructive Technology Assessment. *Technological Forecasting and Social Change*, 54: 251–68.

Schudson, M. 2008. The 'Lippmann–Dewey Debate' and the Invention of Walter Lippmann as an Anti-Democrat 1986–1996. *International Journal of Communication*, 2(1): 1031–42.

Schutz, A. 1964. *Collected Papers II: Studies in Social Theory*. The Hague: Martinus Nijhoff.

Seifert, F. 2006. Local Steps in an International Career: A Danish-style Consensus Conference in Austria. *Public Understanding of Science*, 15(1): 73–88

Selinger, Evan, Thompson, Paul and Collins, Harry. 2011. Catastrophe Ethics and Activist Speech: Reflections on Moral Norms, Advocacy, and Technical Judgment. *Metaphilosophy*, 42(1/2): 118–44.

Shapin, Steven. 1979. The Politics of Observation: Cerebral Anatomy and Social Interests in the Edinburgh Phrenology Disputes. *The Sociological Review*, 27: 139–78.

Shapin, Steven. 1988. Understanding the Merton Thesis. *Isis*, 79(4): 594–605.

Shapin, Steven. 1994. *A Social History of Truth: Civility and Science in Seventeenth-Century England*. Chicago, IL: University of Chicago Press.

Sheldon, T. 2009. Dutch Public Health Experts Refute Claims that Human Papillomavirus Vaccination Has Health Risks. *BMJ*: 338.

Sorgner, Helene. 2016. Challenging Expertise: Paul Feyerabend vs. Harry Collins & Robert Evans on Democracy, Public Participation and Scientific Authority. *Reappraising Feyerabend*: Special Issue of *Studies in History and Philosophy of Science Part A*, 57(June): 114–20. doi:10.1016/j.shpsa.2015.11.006.

Stark, J. 1938. The Pragmatic and the Dogmatic Spirit in Physics. *Nature*, 141(April 30): 770–2.

Taleb, N. N. 2010. *The Black Swan: The Impact of the Highly Improbable Fragility*. New York: Random House.

Turner, Stephen P. 2003. *Liberal Democracy 3.0: Civil Society in an Age of Experts*. London and Thousand Oaks, CA: Sage.

Wakeford, T. 2002. Citizen's Juries: A Radical Alternative for Social Research. *Social Research Update*, 37: 1–5.

Wehrens, Rik. 2014. The Potential of the Imitation Game Method in Exploring Healthcare Professionals' Understanding of the Lived

Experiences and Practical Challenges of Chronically Ill Patients. *Health Care Analysis*, 23(3): 253–71.

Weinel, Martin. 2007. Primary Source Knowledge and Technical Decision-making: Mbeki and the AZT Debate. *Studies in History and Philosophy of Science*, 38(4): 748–60.

Weinel, Martin. 2008. Counterfeit Scientific Controversies in Science Policy Contexts. Cardiff School of Social Sciences. www.cardiff.ac.uk/socsi/research/publications/workingpapers/paper-120.html.

Weinel, Martin. 2010. Technological Decision-making under Scientific Uncertainty: Preventing Mother-to-child Transmission of HIV in South Africa. Unpublished Ph.D. thesis, Cardiff University.

Welsh, Ian. 2000. *Mobilising Modernity: The Nuclear Moment*. London: Routledge.

Wenar, Leif. 2013. John Rawls, in Edward N. Zalta (ed.), *The Stanford Encyclopedia of Philosophy*, Winter 2013 edition. http://plato.stanford.edu/archives/win2013/entries/rawls.

Wetmore, Jameson M. 2015. Delegating to the Automobile: Experimenting with Automotive Restraints in the 1970s. *Technology and Culture*, 56(2): 440–63.

Wickson, F. and Wynne, B. 2012. The Anglerfish Deception. *EMBO Reports*, 13(2): 100–5.

Wildavsky, Aron. 1979. *Speaking Truth to Power*. Bosto, MA: Little, Brown.

Wilsden, James and Willis, Rebecca. 2004. *See-through Science: Why Public Engagement Needs to Move Upstream*. London: DEMOS, Green Alliance, RSA and Environment Agency.

Winch, Peter G. 1958. *The Idea of a Social Science*. London: Routledge and Kegan Paul.

Wittgenstein, L. 1953. *Philosophical Investigations*. Oxford: Blackwell.

Wynne, Brian. 1982. *Rationality and Ritual: The Windscale Inquiry and Nuclear Decisions in Britain*. Chalfont St Giles, Bucks.: British Society for the History of Science.

Wynne, Brian 1989. Establishing the Rules of Laws: Constructing Expert Authority, in R. Smith and B. Wynne (eds), *Expert Evidence: Interpreting Science in the Law*. London and New York: Routledge, 23–55.

Wynne, Brian. 1992a. Misunderstood Misunderstanding: Social Identities and Public Uptake of Science. *Public Understanding of Science*, 1(3): 281–304. doi: 10.1088/0963–6625/1/3/004.

Wynne, Brian. 1992b. Risk and Social Learning: Reification to Engagement, in Sheldon Krimsky and Dominic Golding (eds.), *Social Theories of Risk*. Westport, CT: Praeger Publishers, 275–97.

Wynne, Brian. 2003. Seasick on the Third Wave? Subverting the Hegemony of Propositionalism. *Social Studies of Science*, 33(3) (June): 401–18.

Wynne, Brian. 2008. Elephants in the Rooms where Publics Encounter 'Science'? A Response to Darrin Durant, 'Accounting for Expertise: Wynne and the Autonomy of the Lay Public'. *Public Understanding of Science*, 17(1): 21–33. doi: 10.1177/0963662507085162.

Yearley, Steven. 2000. Making Systematic Sense of Public Discontents with Expert Knowledge: Two Analytical Approaches and a Case Study. *Public Understanding of Science*, 9:. 105–22.

Yinger, J. M. 1982. *Countercultures: The Promise and the Peril of a World Turned Upside Down*. New York: Free Press.

Zurita, L. 2006. Consensus Conference Method in Environmental Issues: Relevance and Strengths. *Land Use Policy*, 23(1): 18–25.

Index